AF715288

The Unwritten Code

The Harmony Codex, Volume 1

Sandeep Chavan

Published by Gyrus Vision, 2025.

While every precaution has been taken in the preparation of this book, the publisher assumes no responsibility for errors or omissions, or for damages resulting from the use of the information contained herein.

THE UNWRITTEN CODE

First edition. November 29, 2025.

Copyright © 2025 Sandeep Chavan.

ISBN: 979-8232093075

Written by Sandeep Chavan.

Table of Contents

To the seekers who live between equations and epiphanies—May your pursuit of truth always remain unwritten, evolving, and fearless.

"You seek proof—but what if proof comes only after belief?"

— The Acharya

Opening Note

This book marks the beginning of a bold intellectual journey—one that dares to question the boundaries we have drawn between science and ancient wisdom, between what is measurable and what is meaningful.

In a world where scientific authority and spiritual intuition often sit on opposite ends of the spectrum, *The Unwritten Code* attempts to build a bridge—not to force convergence, but to enable dialogue. This book is not about proving that ancient scriptures are scientific, nor is it an attempt to mystify science with poetic allegory. Rather, it is a search for deeper understanding—a pursuit of the questions that modern inquiry has avoided, forgotten, or underestimated.

At the heart of this work lies a simple premise: *truth does not belong to one tradition, one discipline, or one era.* Whether discovered through experiments in a lab or realized through deep inner reflection, knowledge—real knowledge—must be examined, refined, and integrated.

This is not the end of a search, but the beginning of a new way of asking questions. What if consciousness is not a byproduct but a participant? What if the subtle order described in Vedic thought reflects structured energy dynamics that modern physics is only beginning to sense? What if the limitations of science are not its methods, but its boundaries?

Each chapter in this book is a step across that boundary. A conversation between a rational mind and a timeless intuition. A reflection of the ongoing evolution of human understanding.

As the first in a series that explores this convergence, *The Unwritten Code* lays the foundation. Future volumes will delve deeper into consciousness, energy, time, perception, and the nature of existence itself—continuing the work of decoding what the ancients may have seen, and what science is now beginning to glimpse.

Let this not be a book you read and put down. Let it be a *mirror, a spark, and a challenge*. To think wider. To ask better. To accept that *truth may not be written—but it is not unknowable.*

— Sandeep Chavan

The Harmony Codex

Ancient Indian Philosophy Reimagined

About the Series

This book is part of an ongoing series that reinterprets the foundational ideas of ancient Indian philosophy—not through rituals or belief systems, but through universal laws of balance, consequence, and conscious living.

The series invites readers on a journey across twenty volumes, each one designed as a standalone reflection while contributing to a larger philosophical mosaic. It begins with accessible themes like patterns, ethics, choices, relationships, and leadership. Gradually, it transitions into deeper exploration of mind, energy, time, illusion, consciousness, and universal alignment.

Behind this unfolding vision is a modern philosophical framework called **Universal Energy Dynamics (UED)**—a ripple-based understanding of reality where every action, thought, or intention carries a consequence, echoing through inner and outer fields of life.

While this early volume may not use UED's terminology explicitly, its essence is already embedded:

- **Dharma** as the principle of natural balance

- **Karma** as the dynamic law of consequence
- **Awareness** as the compass of aligned decision-making

Each book subtly prepares the ground for the next, allowing readers to grow alongside the ideas. From Book 4 onward, the language begins to evolve—introducing core UED concepts such as ripple alignment, field sensitivity, structured awareness, and the science of consequence.

This series is not just a reading experience—it's a reawakening of ancient clarity in modern form.

The First Five Books at a Glance:

1. The Unwritten Code — Understanding the invisible patterns that govern human behavior and social design. *(This Book)*

2. The Silent Algorithm — Exploring how natural laws and inner alignment shape our choices and outcomes.

3. The Equation of Life — Decoding Dharma as a law of balance, not belief.

4. The Mind's Blueprint *(upcoming)* — Entering the ripple architecture of thought, identity, and intelligence.

5. Beyond Time and Self *(upcoming)* — Unraveling perception, illusion, and the hidden mechanisms of causality.

A Living Philosophy

This series is for thinkers, seekers, doers, and skeptics alike—anyone willing to pause, reflect, and realign. It is not a return to the past, but a conscious integration of the timeless with the emerging.

You are not just following a philosophy. You are participating in its re-creation.

Preface

For centuries, science and ancient wisdom have been seen as two opposing forces—one driven by empirical evidence, the other by experiential insight. But what if this division is artificial? What if, instead of being opposing perspectives, they are simply different ways of exploring the same fundamental truths?

"The Unwritten Code: Exploring the Science Behind Ancient Knowledge" was born from this question. It is not a book that seeks to prove ancient texts as scientific theories, nor does it attempt to force modern physics into a spiritual framework. Instead, it asks:

- Have we limited our understanding of reality by restricting science to what is already measurable?
- Do ancient descriptions of consciousness, equilibrium, and universal order hint at principles that science has yet to quantify?
- If structured energy and self-regulating systems exist at fundamental levels, what role does observation—and perhaps consciousness—play in them?

This book follows the intellectual journey of Dr. Kartik Varun, a physicist who, despite his initial skepticism, finds himself confronted with evidence that challenges the rigid boundaries of conventional science. Through experiments, debates, and encounters with both supporters and skeptics, he realizes that the real challenge is not proving one worldview over another, but expanding the way science asks its questions.

At its core, this book is about breaking barriers—between science and philosophy, between materialism and consciousness, between what is known and what is yet to be explored.

Who This Book Is For

- For the scientist: This book encourages you to question not just reality, but the very way we define what is real.
- For the seeker: It provides a bridge between the intuitive wisdom of the past and the empirical rigor of modern inquiry.
- For the skeptic: It does not demand belief, only an openness to exploring what has long been ignored.

The Purpose of This Work

We stand at the edge of a new era of discovery. Science has mapped the physical universe, but what lies beyond the observable? If consciousness is more than just a byproduct of the brain, if structured energy equilibrium hints at deeper principles, then perhaps the next revolution in science will not come from deeper equations—but from a broader way of thinking.

This book does not claim to have all the answers. Instead, it offers a question: What if the greatest truths are not those we have already written, but those we have yet to discover?

Welcome to The Unwritten Code.

Sandeep Chavan

How to Read This Book

The Unwritten Code is not a textbook. It is not a religious commentary. It is a journey—through questions, paradoxes, and insights that challenge the rigid borders of modern knowledge. To engage with this book meaningfully, you must not read it as a passive consumer of information, but as an active participant in inquiry.

This book is best read with curiosity and openness. While each chapter builds upon the previous one, you are encouraged to pause, reflect, and even disagree. The purpose is not to convert, but to catalyze new thinking.

You will encounter both scientific references and ancient philosophical ideas. Some may appear conflicting at first. That is intentional. These tensions are not contradictions—they are doorways to deeper understanding. Don't rush to resolve them. Sit with them. Let them provoke thought.

The story follows Kartik, a rational scientist, as he moves from skepticism to exploration, navigating Vedic concepts and modern physics alike. His path mirrors the reader's own shift—from questioning facts to questioning frameworks.

Along the way, you will find:

- Philosophical provocations grounded in ancient texts
- Scientific analogies and models to frame new interpretations
- Conversations that blur the lines between logic and experience

You don't need prior expertise in either science or spirituality. What you need is a willingness to unlearn. To think beyond binaries. And to accept that the unknown is not a void—it is a frontier.

If at any point this book makes you uncomfortable, good. That means it's working.

Read slowly. Question deeply. And remember: *You are not reading to find an answer. You are reading to rediscover the courage to ask better questions.*

Prologue

The auditorium was dimly lit, yet the weight of expectation filled the space more than the glow of the overhead spotlights. Dr. Kartik Varun stood at the podium, scanning the sea of faces before him—scientists, philosophers, scholars, each waiting, some with curiosity, others with skepticism. The hum of whispered conversations faded as silence took its place, thick and anticipatory.

He exhaled slowly. This moment had been inevitable, yet he had not foreseen it. Not years ago, when he had dismissed these very ideas as remnants of an outdated worldview. Not even when he had first begun questioning the limits of his knowledge.

Now, here he was.

"Ladies and gentlemen," he began, his voice even but carrying the weight of something deeply personal, "I have stood on the firm ground of science all my life. I have built my understanding of the universe through equations, experiments, and logic. But somewhere along the way, I realized—there were questions my discipline could not yet answer."

His fingers rested lightly on the podium. He wasn't reading from notes; this was not a lecture. It was an admission.

"There was a time I believed that if something could not be measured, it did not exist. That if it could not be tested, it held no value. And yet, I find myself here, not just as a physicist, but as a seeker—because I have come to understand that knowledge is not confined to one language, one discipline, or one era."

A murmur passed through the audience. He let it settle.

"In our pursuit of understanding the cosmos, we often assume that we are the first to ask these questions. But long before modern science, there were those who sought to understand the nature of existence—not through telescopes or particle accelerators, but through introspection, observation, and reasoning of a different kind."

Behind him, the screen illuminated with images—ancient Sanskrit texts, cosmic maps drawn centuries ago, mathematical structures that bore an uncanny resemblance to modern physics models. Kartik turned briefly to glance at them before facing the audience again.

"The Vedas, the Upanishads—these are not just spiritual texts. They are inquiries. They ask the same questions we do: What is reality? What is time? What is consciousness? And in many ways, they offer perspectives that resonate, sometimes unsettlingly, with what we are only now beginning to uncover."

He paused, allowing the weight of his words to settle. There was no need for grand declarations. The evidence—the undeniable patterns—spoke for itself.

"Today, I do not stand here to argue faith against science," he continued, his voice firm. "Nor do I suggest replacing one with the other. Instead, I offer a simple idea—that perhaps, when we strip away the language, the centuries, the interpretations, we might find that these two seemingly opposite worlds are not in conflict at all."

A beat of silence. He could see the shifting expressions—some intrigued, some wary. That was expected.

"I propose an integration—one that honors the empirical rigor of science while recognizing the depth of wisdom in ancient thought. Not as opposing forces, but as two perspectives looking at the same reality from different vantage points. What we consider myth may be metaphor. What we call science may be a rediscovery."

He let his words linger before stepping back slightly.

"This journey is not mine alone. It belongs to all of us who seek to understand—not just what the universe is, but what it means."

The room remained still, the silence no longer just anticipation, but contemplation.

And in that moment, Kartik knew—this was no longer a hypothesis.

It was the beginning of a paradigm shift.

Part I — The Rational Foundation

A Scientist's Restlessness

From scientific certainty to intellectual curiosity

Chapter1: The Skeptic's Threshold

Dr. Kartik Varun had always believed that every phenomenon could be dissected, measured, and explained through the precision of empirical science. His world revolved around equations and observable facts, leaving no space for the intangible or the mystical. He had spent years building a reputation as a physicist who sought truth in logic, not in ancient scriptures or philosophical musings. But life, indifferent to his certainties, had other plans.

The news of his uncle's passing had brought him back to his ancestral home—a place steeped in traditions he had long dismissed as relics of a bygone era. What was meant to be a brief return, a formality to fulfill familial obligations, soon became something more. Seated among mourners, he found himself part of a ritual he neither understood nor valued—a Pravachan, a spiritual discourse intended to offer solace and wisdom in the face of loss. Kartik had expected vague platitudes, mythological embellishments, and sentimental reflections on the impermanence of life.

Instead, he found himself confronted with something entirely different.

The speaker's words carried an unexpected depth, an eloquence that resonated beyond the realm of faith and belief. As the discourse unfolded, Kartik recognized patterns—ideas that bore striking similarities to concepts he had encountered in quantum physics and cosmology. It was unsettling. The boundaries between the spiritual and the scientific, which he had always regarded as separate and irreconcilable, began to blur.

What started as an hour of reluctant endurance became a moment of revelation. Kartik left the Pravachan with a mind restless with questions, unable to dismiss the possibility that within these ancient words lay a wisdom he had overlooked. What if the Vedic philosophers, in their poetic verses and allegories, had glimpsed truths that modern science was only beginning to articulate?

This thought, once unthinkable, now refused to leave him. And so, against everything he had once believed, Kartik found himself standing at the threshold of something unexpected—an intersection where the past and the future, the mystical and the empirical, were not at odds, but perhaps, inextricably linked.

A Reluctant Return

The phone call came late in the evening, its solemn tone cutting through the hum of his laboratory's fluorescent lights. Dr. Kartik Varun was immersed in his research, his mind weaving through complex equations on the quantum nature of time, when his mother's voice pulled him back to reality.

"Kartik... your uncle has passed away. The funeral is tomorrow."

He exhaled, his fingers gripping the pen a little tighter. The weight of familial duty settled upon him, clashing with his instinct to retreat into the comfort of logic and routine. It had been years since he had returned home, not out of estrangement, but because his world had gradually shifted away from the one he had left behind.

"I understand, Maa. I'll take the first flight," he replied, his voice measured.

Hanging up, Kartik leaned back in his chair, staring at the half-finished calculations on his screen. There was no place for emotions in physics, just as there was no place for rituals in reason. Yet, here he was—returning to a home that still echoed with traditions he had spent a lifetime distancing himself from.

Homecoming Amidst Grief

The moment Kartik stepped into his ancestral home, he was engulfed in a world that felt both familiar and distant. The air was thick with the scent of incense, mingling with the muted murmurs of gathered relatives. The house, with its high wooden beams and flickering oil lamps, was frozen in time—a stark contrast to the sterile, modern laboratories he was accustomed to.

His mother embraced him briefly, her eyes tired yet relieved. Around him, cousins and distant relatives exchanged quiet condolences, their expressions filled with a grief he struggled to emulate.

"Sit, beta," his mother said, motioning to a chair near the main hall where mourners had gathered. *"The priest will begin the rituals soon."*

Kartik hesitated before sitting, his gaze flickering to the framed photograph of his uncle, adorned with marigold garlands. A man of strong beliefs and an unwavering faith in tradition, his uncle had often chided him for his dismissive attitude toward spirituality.

Kartik could almost hear his voice now—*Science will not explain everything, Kartik. One day, you'll see.*

A flicker of guilt passed through him. He respected his uncle, but the rituals that followed death? They had always seemed redundant to him, a means of comforting the living rather than serving the departed.

The Scientist's Discomfort

As the ceremony progressed, Kartik's unease grew. Sanskrit chants filled the room, their meanings lost on him despite their rhythmic, almost hypnotic cadence. The priest, clad in simple saffron robes, moved with practiced grace, offering prayers and symbolic offerings to the sacred fire.

Kartik observed it all with the detached gaze of a scientist, noting the precision of the rituals, the repetitive nature of the hymns. There was structure, certainly—but was there meaning?

He shifted uncomfortably, his rational mind dissecting the process. Could these rituals genuinely guide a soul to liberation, as they claimed? Or were they remnants of an age where knowledge was woven into myth and metaphor, preserved through generations simply because no one dared to challenge them?

His thoughts were interrupted by his cousin, who leaned in and whispered, *"There's a Pravachan this evening, after the ceremonies. It's in honor of Kaka."*

Kartik barely managed to suppress a sigh. He knew what that meant—a discourse on life, death, and spiritual wisdom, led by a learned speaker. He had no patience for it, yet declining would seem insensitive.

"Who's the speaker?" he asked out of courtesy.

"Acharya Vishwanath," his cousin replied. *"He's well-respected, has studied ancient scriptures deeply. People say he has a way of explaining things in a way even modern minds can appreciate."*

Kartik forced a nod. He had spent his life dismissing such things as poetic interpretations of an uncertain world. But he also knew himself well enough to recognize that, beneath his skepticism, there was a flicker of curiosity.

And that was a dangerous thing.

Return to a World Left Behind

As the car pulled up to his ancestral home, Kartik felt an inexplicable weight settle on his chest. The house stood just as he had left it years ago—its whitewashed walls weathered by time, the wooden doors adorned with intricate carvings of deities and motifs he barely remembered. The scent of damp earth mixed with the lingering aroma of incense, a fragrance so deeply woven into his childhood that, for a brief moment, he could almost believe he had never left.

But he had.

Stepping out of the car, he adjusted the strap of his laptop bag—a silent reminder of the world he had come from. He wasn't the young boy who once ran barefoot through these halls, nor the teenager who had argued with elders about the "scientific absurdities" of their beliefs. He was Dr. Kartik Varun now—a man of logic, of reason, of carefully constructed proofs. And yet, standing at the threshold of his past, he couldn't deny the subtle unease that stirred within him.

The front door creaked open before he could knock. His mother stood there, draped in a simple white saree, her face lined with grief and fatigue. For a fleeting second, she said nothing, merely looking at him with the quiet relief of a mother whose child had come home, no matter how many years had passed.

"You came," she said, her voice soft but resolute.

Kartik nodded. *"Of course, Maa."*

She stepped aside, allowing him in. The house was filled with hushed conversations, the occasional rustling of fabric as relatives shifted uncomfortably in their seats. Mourners sat in the main hall, their heads bowed, some whispering prayers while others simply stared ahead, lost in thought. He recognized many faces—cousins, uncles, aunts—but they seemed like remnants of another life, one he had neatly compartmentalized.

His mother gestured toward a small brass lamp burning near a framed photograph of his uncle, now adorned with marigold garlands. Kartik swallowed hard. His uncle had been a man of conviction, deeply rooted in tradition, yet always kind. He had never chastised Kartik for his skepticism but had often smiled knowingly whenever Kartik dismissed spirituality as primitive storytelling.

He had once told him, *"One day, you will see that the universe is not only what you can measure, but also what you can feel."*

Back then, Kartik had scoffed at the notion. Now, as he looked at the photograph, he felt an unfamiliar pang—was it grief, or something deeper, something unsettling?

His thoughts were interrupted as a cousin approached. *"Kartik bhaiya, Acharya Vishwanath is coming this evening for the Pravachan. It's for Kaka's peace. You will attend, right?"*

Kartik hesitated, the old resistance rising in him like muscle memory. He had sat through many Pravachans as a child, listening but never absorbing, dismissing them as poetic ramblings meant to comfort those who feared death.

But this time, something was different. This time, he wasn't just a boy forced to listen—he was a scientist who had spent his life dissecting reality. And for reasons he couldn't quite explain, he felt an odd compulsion to sit through this one.

Would it be the same predictable sermon? Or was there something in those ancient words that he had overlooked all these years?

He exhaled. *"I'll be there."*

Even as he said it, he wasn't sure why.

The Scientist Out of Place

Kartik sat cross-legged on the floor, his back stiff, the dull murmur of Sanskrit chants echoing through the room. The flickering glow of oil lamps cast elongated shadows on the walls, their light reflecting off the brass vessels arranged meticulously before the priest. Each item—sandalwood paste, rice grains, betel leaves—was placed with ritualistic precision, their significance lost on him.

He had spent years in laboratories where precision had a purpose, where calculations had to be exact, where every equation was a bridge between hypothesis and reality. But here, amidst ancient customs and whispered prayers, precision seemed ornamental—an elaborate performance steeped in belief rather than necessity.

The priest continued his rhythmic incantations, his voice rising and falling like waves against the shore. Kartik's eyes wandered over the room. His mother sat with quiet reverence, her hands folded, her gaze lowered. His cousins mirrored the gestures of the elders, performing each ritual as if by instinct. None of them seemed to question why or how. They simply believed.

And that was what unsettled him most.

Science had taught him to question, to demand evidence before acceptance. But here, faith required no proof—only participation. He was an outsider, a visitor in a world where certainty was not derived from data but from devotion.

A deep sigh escaped him. He knew his presence was expected, yet he felt no connection to what was unfolding before him. The priest sprinkled water in measured arcs, muttering prayers Kartik did not understand. Each action, each chant, carried the weight of tradition, yet to him, it was a language without meaning.

His fingers tapped absently against his knee—a habit born from years of restless thinking. *How much of this is symbolism? How much of it is blind adherence?*

His thoughts were interrupted when the priest motioned for him to extend his hands. Hesitantly, Kartik complied, his palms facing upward. A small mound of rice was placed in them, followed by drops of water. The priest instructed him to repeat a phrase in Sanskrit. The words felt foreign on his tongue, unnatural.

"This offering is made in remembrance of your uncle," the priest explained, his voice patient but firm.

Kartik hesitated. An offering. To whom? To what? His uncle was gone. Science told him that death was the irreversible cessation of biological function. No ritual, no prayer could alter that fact. So why did everyone here act as if these gestures mattered?

He glanced at his mother. She did not meet his gaze, but he could see the silent plea in her posture—*Just do this, Kartik. For him. For us.*

Suppressing the urge to sigh, he followed through with the motions, murmuring the words as instructed. A ripple of satisfaction passed through the room, the subtle approval of those around him.

He felt no different.

No revelation. No sudden sense of peace. Only the quiet confirmation that, once again, he did not belong.

As the rituals continued, his thoughts drifted. He longed for the sterile certainty of his equations, for the logic of the universe that followed laws, not beliefs. But deep within, a quiet question lingered—*If everything must be measurable to be real, why does belief persist across centuries?*

It was a question he had always dismissed before.

But tonight, it refused to be ignored.

A Ritual, A Reluctant Observer

The smoke from the sacred fire curled toward the ceiling, its tendrils weaving through the dimly lit hall, casting shifting patterns of light and shadow. Kartik watched in silence, his fingers interlocked, his mind far removed from the solemnity that surrounded him. The rhythmic cadence of the priest's chants continued, their syllables flowing in a language that, despite his heritage, felt foreign to him.

Around him, his family sat in quiet reverence—his mother whispering prayers under her breath, his cousins bowing their heads with unquestioning devotion. There was a synchrony in their movements, an unspoken understanding of every ritual, every gesture. It was a world where belief was inherited like an heirloom, passed down from one generation to the next without challenge.

Kartik felt no such inheritance.

He had spent years conditioning himself to resist the pull of tradition, to seek answers only in what could be observed, tested, and proven. Yet here he was, an unwilling participant in a ceremony that seemed to demand not just presence, but faith.

The priest flicked droplets of water into the fire, his voice steady as he recited verses Kartik did not comprehend. Every motion had a meaning, every offering a purpose—at least to those who believed. But to him, it was all just a sequence of actions performed out of obligation rather than understanding.

A distant cousin leaned in and whispered, *"Acharya Vishwanath is coming this evening for the Pravachan."*

Kartik turned slightly, his brows drawing together. *"Pravachan?"*

The cousin nodded. *"A discourse. It's for Kaka's peace. And for all of us—to find meaning in his passing."*

Another sermon. Another attempt to find cosmic significance in death. Kartik had heard many in his childhood, each one filled with poetic musings on the nature of life and the soul's journey beyond. He had always dismissed them as comforting fictions, stories woven to soften the finality of loss.

"*You should stay,*" his cousin added, as if sensing his reluctance. *"Acharya Vishwanath is not like the others. People say he makes sense—even to those who don't believe."*

Kartik remained silent, his gaze fixed on the fire. He had no intention of being swayed by a spiritual lecture, yet something about the way his cousin had phrased it unsettled him.

A speaker who made sense even to skeptics?

He wasn't sure why, but that thought lingered longer than he expected.

The Gathering at Dusk

As the last rays of sunlight faded into the evening sky, the hall was bathed in the soft glow of oil lamps, their flickering light casting elongated shadows on the walls. The scent of incense hung thick in the air, its presence both soothing and overpowering. Mourners settled into quiet expectation, their whispers fading into silence as they turned their attention to the elevated wooden dais at the front of the room.

Kartik sat near the edge of the gathering, his posture rigid, his mind still resisting the moment. He had attended such gatherings as a child, always watching from the periphery, unwilling to surrender to what he had dismissed as elaborate storytelling. And yet, tonight felt different. Perhaps it was the weight of his uncle's absence, or perhaps it was the slow realization that his detachment did nothing to change the reality of the moment.

A hush fell over the audience as Acharya Vishwanath took his place on the dais. Dressed in simple saffron robes, his presence was one of quiet command rather than ostentatious authority. He did not carry the air of a preacher seeking reverence, nor did he appear burdened by the expectations of the gathering. Instead, he exuded a calmness that seemed to stretch beyond the confines of the room, as if he was merely a conduit for something much older, something that had existed long before any of them had gathered here.

Kartik studied him with a detached curiosity. The Acharya's eyes moved over the assembly, acknowledging each presence with a slow, deliberate gaze. He did not begin with a chant or a customary invocation. Instead, he simply sat in silence, letting the weight of the moment settle upon the audience.

It was an unusual approach, one that made Kartik shift slightly in his seat. Most speakers filled the air with words, eager to establish authority. But this man allowed the silence to stretch, unbroken and undisturbed.

And then, in a voice both deep and measured, he spoke.

"Grief is a teacher."

His words were not dramatic. They were not laden with poetic embellishment. Yet, they carried the weight of something undeniable, something that resonated with the unspoken emotions in the room.

Kartik felt the shift in the air—the subtle but unmistakable change in the way people listened.

And despite himself, he found that he was listening too.

An Outsider Among the Faithful

Kartik sat with his arms loosely crossed, his back straight, his expression unreadable. The Pravachan had begun, and yet, he remained an observer rather than a participant. Around him, heads were bowed in silent contemplation, faces softened by the weight of reflection. For them, these words were a source of comfort, a guiding light in a moment of grief. But to him, they were echoes of a past he had spent years distancing himself from.

The Acharya's voice carried across the hall, neither too loud nor too soft, each syllable precise, deliberate. *"Grief is a teacher,"* he had begun, and now, as he spoke of life's impermanence, of the ebb and flow of existence, Kartik listened—not with devotion, but with a scientist's detached curiosity.

What makes people so certain? he wondered, glancing at his mother. Her hands rested gently in her lap, her eyes half-closed, as if she was absorbing every word, letting it settle into the deepest corners of her being.

The Acharya continued, his tone neither forceful nor pleading. *"We mourn because we believe something has been taken from us. But what if loss is not an ending? What if it is simply a change—one that we are not yet ready to accept?"*

Kartik exhaled slowly. *Change is measurable. It has equations, formulas, and explanations. But this—this is sentiment wrapped in metaphor.*

Yet, as he scanned the faces around him, he saw not blind faith, but something else—acceptance. The kind that did not need proof, did not demand explanation. For them, belief was enough.

His gaze returned to the Acharya, who seemed to pause, as if sensing the presence of those who were still unconvinced.

"And for those who question," the Acharya said, his voice calm yet unwavering, *"for those who seek proof before belief—perhaps the lesson is not in accepting without reason, but in asking the right questions."*

Kartik's fingers twitched slightly. It was as though the words had been spoken directly to him.

Coincidence? Or had the Acharya already sensed his doubt?

For the first time that evening, Kartik felt something stir—not belief, not yet, but a question.

And questions, he knew, had a way of leading to unexpected answers.

The Voice That Commands Attention

Acharya Vishwanath spoke with an unhurried cadence, his voice steady, carrying the weight of something deeper than mere knowledge—an understanding that felt lived, not just learned. Unlike the spiritual discourses Kartik had endured in his youth, there was no flourish, no grand theatrics to demand attention. And yet, everyone listened.

Kartik had expected the usual metaphors—scriptures woven into comforting narratives, words wrapped in poetic ambiguity. But the Acharya was doing something different. He was not reciting wisdom as a sermon; he was peeling away layers, exposing truths that seemed to exist just beneath the surface of perception.

"We look at the world and see only movement, change, cause and effect," the Acharya said. *"We assume that what we observe is all there is. But what if we are only perceiving a fraction of reality? What if the truth is woven in spaces we do not yet know how to measure?"*

Kartik frowned slightly. It was an interesting choice of words. He had encountered similar ideas in physics—not in philosophy, but in quantum mechanics, where the nature of reality itself seemed uncertain. Where particles existed in multiple states, where observation changed outcomes. He had spent years questioning the fundamental nature of existence, but never in this language.

The Acharya continued. *"We often believe that knowledge is only what can be seen, what can be counted. But ancient wisdom suggests otherwise. It speaks of forces beyond visibility, of energies not yet understood, of truths that do not require validation to be real."*

There it was—that quiet challenge. The assertion that not everything needed proof. Kartik stiffened slightly, instinctively resisting the implication. *Without proof, there is only assumption.*

And yet...

Hadn't modern science struggled with the same limitations? Hadn't physics, in its most advanced form, reached the edge of what could be seen and measured? The universe itself held mysteries beyond human comprehension, questions that could not yet be answered.

Kartik's mind wrestled with the thought, but his body betrayed him. He had leaned forward ever so slightly, his resistance faltering in the face of something he could not yet name.

The Acharya's gaze swept the hall, and for the briefest moment, Kartik felt as if their eyes met—as if the man could see the turmoil within him, the battle between logic and the quiet whisper of curiosity.

And then the Acharya spoke again.

"Some of you are here to find peace," he said, his voice softer now. *"Some of you are here because you believe. And some... are here only to listen, unsure of what they will find."*

Kartik exhaled slowly.

He had come as an observer.

But somehow, without realizing it, he had begun to listen.

Logic Confronts the Unexplained

Kartik had come prepared to endure the discourse, not engage with it. He had assumed it would be a collection of allegories designed to comfort those who sought meaning in loss, a ritual of words rather than reason. But now, as the Acharya spoke, he found himself gripping the edge of his seat—not in awe, not yet, but in discomfort.

Because what he was hearing did not sound like blind faith. It sounded like science.

"The universe is not chaotic," the Acharya continued. *"It appears so only to the untrained eye, to the mind that sees events in isolation. But look deeper, and you will find patterns—an order woven into the very fabric of existence. That which we call Dharma is not a rule imposed from above; it is the balance that sustains all things."*

Kartik's breath slowed. *Patterns. Order. Balance.* These were not spiritual abstractions; they were the foundation of physics. From the laws of motion to the fundamental forces of nature, everything he had studied suggested that the universe was governed by unseen principles, invisible frameworks that dictated how energy moved, how matter interacted.

The Acharya's voice carried on, undisturbed by the silent ripples of doubt and intrigue it was stirring in Kartik's mind. *"We often assume that what cannot be measured does not exist. But tell me, if something moves without being seen, if it exerts influence without being touched—does it not still shape the world?"*

Kartik's pulse quickened. That was *exactly* the question physics had been grappling with for decades. Quantum mechanics was built on such paradoxes—particles that existed in multiple places at once, forces that operated across impossible distances, energy that behaved like both a wave and a particle depending on whether it was being observed.

These were modern discoveries. And yet, here was a man, dressed in simple robes, speaking of them as if they had been known for millennia.

The scientist in him resisted. Ancient texts did not contain quantum equations. They did not predict wave function collapses or the probabilistic nature of particles. But then... had science not spent centuries dismissing ideas, only to later confirm them?

He thought of how Newton's rigid mechanical universe had given way to Einstein's relativity, how even Einstein had struggled to accept the eerie uncertainty of quantum mechanics. How much of what modern science now accepted had once been considered absurd?

What if these philosophies were not superstition, but an earlier way of describing the same truths?

The thought unsettled him. He did not believe in mysticism. He did not accept knowledge without evidence.

But he could not ignore the possibility that what he was hearing—words spoken thousands of years before physics had even defined its first principles—aligned too closely with ideas he had spent his life studying.

His skepticism did not vanish. But it wavered.

For the first time, he wasn't sure if he was resisting because the ideas were flawed—or because they forced him to question the certainty he had built his life upon.

A Spark in the Mind

Kartik leaned forward, his fingers interlocked, his mind caught between skepticism and an unfamiliar sense of intrigue. The Acharya's words had started as philosophical musings, but now, they felt like quiet provocations—direct challenges to the rigid framework of logic he had spent his life constructing.

"We speak of time as if it moves in a straight line," the Acharya continued, his voice steady, deliberate. *"We measure it, divide it into past, present, and future. But what if time is not a river that flows in one direction? What if it is a cycle—an eternal return?"*

A murmur of agreement rippled through the audience, but Kartik barely noticed. His mind raced through equations, theories, and the unresolved debates in physics about the nature of time itself. He had encountered this idea before—not in scripture, but in the paradoxes of relativity and the strange mechanics of the quantum world.

Modern physics had already shaken the classical notion of time. Einstein had shown that time was not absolute, that it stretched and compressed depending on gravity and motion. And in quantum mechanics, there were equations where time did not even have a preferred direction—suggesting that, at its fundamental level, the universe did not experience time the way humans did.

And yet, here was an ancient teacher, echoing the same ideas in an entirely different language.

Kartik exhaled, forcing himself to remain grounded. *Coincidence.* He had seen it before—how people retrofitted ancient beliefs to align with modern discoveries. This was likely no different.

And yet, the questions refused to fade.

The Acharya was still speaking. *"If time moves in cycles, then what we call the past does not disappear—it remains, woven into the present in ways we do not always perceive. Knowledge is not always discovered; sometimes, it is only remembered."*

A quiet hum of thought stirred in Kartik's mind. *Not discovered, but remembered.*

He had read theories that played with similar ideas. The concept of the universe as an endless cycle of expansion and contraction—the 'big bang' not as a singular event, but as one of many, in an infinite loop of creation and dissolution. Some physicists had even proposed models where information was never truly lost, where the past could, in some way, imprint itself onto the future.

It was a hypothesis, unproven, existing only in speculative papers. But it was there.

And now, in this dimly lit room, illuminated only by flickering lamps and the glow of belief in the eyes of those around him, Kartik found himself wondering—was this simply metaphor, or had these ancient thinkers glimpsed something deeper?

For the first time, the line between philosophy and science blurred, not because one was bending to the other, but because they seemed to be speaking the same truth—separated only by the language of their time.

The thought was disorienting. And exhilarating.

Kartik had come here expecting to endure, not to engage. He had assumed this Pravachan would reaffirm his skepticism, not challenge it.

But here he was.

And for the first time, he was listening—not as a skeptic, not as a scientist dissecting belief, but as a mind confronted with a question it could not yet answer.

Standing at the Crossroads

The Acharya's final words lingered in the air long after he had fallen silent. Around Kartik, the gathering stirred, the hush of deep contemplation giving way to the quiet rustle of movement—folded hands lowering, murmured prayers concluding, bodies shifting as people prepared to leave.

But Kartik remained seated.

He should have risen, dusted off the evening, and moved on. That was the logical thing to do—file this experience away as an interesting but ultimately irrelevant intersection between philosophy and science. And yet, his body refused to obey the command of reason.

Something deeper held him still.

His mind replayed fragments of the Pravachan, words that had unsettled him in ways he did not yet fully understand. *Not everything unseen is unreal. Time does not vanish; it transforms. Knowledge is not always discovered; sometimes, it is remembered.*

They were poetic ideas, certainly. But why, then, did they feel so uncomfortably familiar? Why did they remind him of the very questions modern physics had yet to answer?

The scientist in him rebelled. *Correlation is not causation,* he reminded himself. Just because ancient texts spoke of patterns and cosmic balance did not mean they had arrived at these ideas through the same rigorous methods that science demanded. *Intuition is not proof. Faith is not evidence.*

And yet.

Hadn't the greatest scientific discoveries begun as nothing more than absurd questions? Hadn't every breakthrough started with someone daring to look beyond what was already known?

Newton once sat beneath a tree and asked why the apple fell downward. Einstein wondered what would happen if he could chase a beam of light. What if these ancient thinkers had done the same—gazing at the universe, asking different but equally profound questions, and attempting to answer them in the only language they knew?

The possibility gnawed at him.

Had he spent his life dismissing something that was not primitive mysticism, but an early attempt to articulate truths that modern science was only now beginning to grasp?

A soft voice interrupted his thoughts. *"You're troubled."*

Kartik turned, startled. It was the Acharya, standing a few feet away, observing him with a knowing calm. The hall had emptied considerably, but a few lingered in quiet conversation at the edges of the space.

Kartik hesitated before responding. *"Not troubled. Just... thinking."*

The Acharya gave a small smile. *"Thinking is good. Unless it is only to reinforce what one already believes."*

Kartik felt an odd heat rise in his chest. It was a simple statement, yet it exposed the very thing he had been trying to ignore. He had come here expecting to validate his skepticism. Instead, he had left with more questions than answers.

The Acharya continued, his voice as steady as it had been during the discourse. *"Truth does not fear inquiry, Dr. Varun. It is only the mind that resists change."*

Kartik stiffened slightly at the mention of his name. He had not introduced himself.

The Acharya chuckled, as if reading his thoughts. *"Your name was mentioned during the prayers. But even if it wasn't, I would have known you were a man of science. You listened differently."*

Kartik exhaled, rubbing his temples. He felt like a man standing at the edge of a great divide, uncertain of whether to cross.

Finally, he met the Acharya's gaze. *"You speak with certainty. But I need proof."*

The Acharya nodded, as if this answer pleased him. *"Then seek it."* He paused for a moment, his expression unreadable. "And when you do, Dr. Varun, you may find that the questions you ask are older than you think."

The response was not defensive, not dismissive—just an invitation.

Kartik did not realize it then, but that single sentence would stay with him, shaping the path ahead in ways he had yet to understand.

For now, he stood up, adjusting his sleeves. The evening air had cooled, and outside, the night stretched wide and endless.

A universe of questions awaited him.

And for the first time in a long while, he was willing to ask them.

Chapter 2: Echoes of Ancient Wisdom

The night air was crisp as Kartik stepped out of the hall, but his mind was anything but clear. The Pravachan had ended, the crowd had dispersed, yet the weight of the discourse lingered, pressing against the walls of his certainty.

He had come expecting a ritualistic monologue, a series of spiritual platitudes meant to soothe the grieving. Instead, he had been confronted with something disturbingly precise—words that did not merely console but provoked, challenged, and, worst of all, aligned with ideas he had encountered in physics.

That was what unsettled him the most.

He had spent his life separating science from belief, reason from faith. But tonight, those boundaries had blurred, and for the first time, he wasn't sure where one ended and the other began.

As he walked through the quiet courtyard, his mother's voice called from behind. *"Kartik, you're quiet."*

He turned, offering a faint smile. *"Just tired."*

She nodded knowingly. *"The Pravachan can do that. Sometimes, it speaks to us in ways we don't expect."*

He hesitated before responding. *Does it?* He wasn't sure if the Pravachan had spoken to him, or if his own mind had betrayed him—pulling threads from ancient philosophy and weaving them into the equations he understood.

Or perhaps... those threads had always been there.

He exhaled, forcing his thoughts to settle. He would not jump to conclusions. He needed proof.

And for that, he needed to listen again.

A Shift in Perspective

The dim glow of the bedside lamp cast long shadows across the walls of Kartik's childhood room. It had been years since he had slept here, yet everything remained unchanged—the same bookshelves, the same wooden desk, even the faint scent of old paper and incense that seemed woven into the air. But tonight, the familiarity of the space offered no comfort. His mind was far from the present, tangled in the echoes of the Pravachan.

He sat on the edge of the bed, rubbing his temples, trying to rationalize why the Acharya's words had unsettled him so deeply. It wasn't faith that had shaken him—he had long since learned to dismiss religious rhetoric. No, this was different. The discourse had not relied on devotion or blind belief. It had relied on patterns, on principles, on an almost mathematical precision that bore an uncanny resemblance to the very laws he had spent his life studying.

The universe is not chaotic. Look deeper, and you will find patterns.

Kartik exhaled sharply. He had spent years staring at cosmic equations that said the same thing. Whether in the delicate balance of planetary motion or the strange behavior of subatomic particles, order existed beneath what seemed random. But how could an ancient discourse, spoken in a hall filled with grieving people, mirror the equations of modern physics?

He leaned forward, elbows on his knees. *I'm reading too much into it.* He wanted to believe that. And yet, the questions wouldn't let go.

Was it coincidence, or had the Acharya truly articulated something deeper? Had these ancient thinkers arrived at the same realizations—just in a different language, one that modern science had yet to decode?

He stood abruptly, pacing toward the window. The night stretched dark and endless, a sky filled with silent stars. For the first time in years, he looked at them not as a physicist, but as a man caught between two worlds—one built on formulas, the other on something less tangible, yet equally structured.

His own thoughts unsettled him.

Science had been his anchor, his unwavering truth. But tonight, for the first time, it felt like something else was pulling at the edges of his understanding.

Not disproving it.

Not contradicting it.

But expanding it.

And that terrified him more than anything.

Bridging the Unseen and the Known

The next evening, Kartik found himself back in the Pravachan hall, though this time, his presence wasn't out of obligation. He had returned willingly—though he wasn't sure if it was to challenge the Acharya's words or to understand them better.

The hall was just as it had been the night before, the glow of oil lamps flickering against the walls, the air carrying the scent of sandalwood and aged wisdom. Mourners had gathered again, some seeking comfort, others perhaps searching for something deeper. Kartik sat near the back, arms crossed, determined to observe without being drawn in.

The Acharya sat in the same poised manner, his presence calm yet commanding. When he spoke, it wasn't with the weight of a preacher, but with the ease of someone unraveling a truth that had always existed, waiting to be seen.

"The universe," he began, his voice steady, *"is not what we perceive it to be. Our senses offer only a fraction of reality. The Rishis who composed the Vedic texts understood this—they spoke of multiple layers of existence, of energies unseen, of forces that shape what we call reality."*

Kartik's fingers twitched slightly. This wasn't unfamiliar. *Our senses offer only a fraction of reality.* He had read similar arguments in physics papers that explored the limitations of human perception—how visible light was only a sliver of the electromagnetic spectrum, how dimensions beyond the observable three might exist, how space itself wasn't truly empty but teeming with quantum fluctuations.

"We think of existence as bound by what we can touch, see, and measure," the Acharya continued. *"But the ancients told us otherwise. They spoke of the universe as a great web—one in which everything is connected, where action and thought ripple through unseen threads."*

Kartik's mind latched onto a single phrase: a great web.

For centuries, physics had sought to explain the fundamental nature of reality. Newton had given us laws of motion, Einstein had rewritten them with relativity, and now, physicists grappled with the paradoxes of quantum mechanics—where particles could exist in multiple places at once, where they seemed entangled across vast distances, their fates linked in ways that defied logic.

Entanglement.

A web.

Kartik sat up straighter. Was the Acharya, in his own way, describing the interconnected nature of reality—something physics was only now beginning to quantify?

The Acharya's voice softened. *"The Vedic texts tell us that what is seen is shaped by what is unseen. That which moves the stars also moves within us. That which exists beyond time influences what we call the present. We are not separate from the universe, nor is it separate from us."*

Kartik clenched his jaw. The idea was poetic, certainly. But it also echoed something eerily familiar. *Quantum field theory proposed that all particles emerged from underlying fields of energy, that reality was a continuous fabric where separation was an illusion.*

Coincidence?

Or had these ideas, hidden in metaphor and verse, been understood long before equations gave them form?

For the first time, Kartik wasn't merely listening.

He was **wondering.**

The Fabric of Reality

The Acharya's voice carried through the hall, steady and unhurried. *"The cosmos does not move by chance. It follows an order—an eternal rhythm that maintains balance. This order, in the Vedic texts, is called* ***Rta.*** *It governs the movement of stars, the cycles of nature, the harmony within the smallest atom and the vastest galaxies."*

Kartik's fingers pressed lightly against his knee. Order. Balance. Predictability. These were not unfamiliar concepts. Physics, at its core, was the study of order within apparent chaos—the principles that dictated how energy moved, how forces interacted, how the universe sustained itself.

The Acharya continued. *"When Rta is followed, there is harmony. When it is disturbed, disorder spreads. This is not a law imposed by a deity—it is the fundamental structure of existence. It is why the planets do not stray from their paths. It is why seasons change with precision. It is why life sustains itself in delicate equilibrium."*

Kartik's mind raced. Entropy. The second law of thermodynamics stated that systems naturally moved toward disorder, toward greater randomness. And yet, there was always an underlying structure—something resisting complete chaos. Stars burned in precise ways, chemical reactions followed specific sequences, even biological life emerged and thrived through self-regulating mechanisms.

"The ancients observed this balance," the Acharya continued, *"not in equations, but in the cycles of nature, in the rhythm of breath, in the patterns of the night sky. They did not call it thermodynamics or gravitational laws, but they understood that the universe was not random. It moved with purpose."*

Kartik exhaled sharply. This was not blind faith. This was observation.

The Acharya's gaze swept across the hall. *"Some believe that because they do not see this order, it does not exist. But one does not have to see the wind to know it moves the trees."*

Kartik's breath stilled. One does not have to see it to know it exists.

He thought of quantum fields—invisible forces that permeated space, giving rise to particles, energy, and matter itself. He thought of the fabric of spacetime, bending under the weight of massive objects, unseen yet measurable.

Had the ancients glimpsed these truths—not through numbers, but through intuition? Through observation sharpened by a different lens?

Kartik's skepticism remained, but something in him wavered.

What if **Rta** was not just a philosophical concept, but an early articulation of the order that modern physics was still struggling to fully understand?

What if he was no longer looking at two separate worldviews—but the same truth, told in different languages?

The Nature of Time

"The past does not vanish," the Acharya said, his voice steady yet filled with something almost timeless. *"It folds into the present. And the present is not a single moment—it is shaped by the echoes of what came before and the patterns of what is yet to unfold. Time, as we perceive it, is an illusion."*

Kartik sat motionless, his mind caught between the familiarity of the statement and the implications it carried. *Time is an illusion.* He had encountered this idea before—not in philosophy, but in relativity, quantum mechanics, and cosmology.

"In the Vedic view," the Acharya continued, *"time does not move in a straight line. It moves in cycles. Creation and destruction, expansion and contraction—nothing begins nor ends, it only transforms."*

Kartik's fingers curled into his palm. The idea was eerily similar to something modern physics had proposed—the cyclic universe model.

Cosmologists had long debated the nature of time and the fate of the universe. The Big Bang theory suggested the universe had a beginning, an origin point from which everything expanded. But some theories argued that this was not the first creation, nor the last. Instead, the universe expanded, collapsed, and was reborn in endless cycles, a rhythmic breathing of cosmic proportions.

And here was the Acharya, describing the same concept—not in equations, but in verses written millennia ago.

Kartik's mind raced further. Time dilation in relativity showed that time was not absolute; it stretched and compressed depending on motion and gravity. In the quantum world, time didn't seem to flow in a single direction—it was an abstract property, malleable under different conditions.

Could these ancient thinkers have grasped, in essence, what modern physics was still struggling to define? Had they, through observation and introspection, arrived at the understanding that time was not linear, but a continuum, bending and shifting in ways beyond human perception?

Kartik shook his head, trying to ground himself. *Coincidence?*

The Acharya's voice interrupted his thoughts. *"If time were truly linear, it would break. If it moved only forward, it would never return to balance. But just as the sun rises and sets, just as the tides ebb and flow, time follows a rhythm that we do not yet fully grasp."*

Kartik swallowed hard. This was no vague spiritual allegory. It was physics.

Not in the way he had learned it—but in the way it had existed long before he ever questioned it.

And for the first time, he wasn't sure which version was more accurate.

The Observer's Paradox

The Acharya paused for a moment, letting the weight of his last statement settle over the gathering. Then, with a quiet certainty, he said, *"Reality is not fixed. It shifts, not just by its own nature, but by the presence of the one who perceives it."*

Kartik sat up straighter.

He had expected discussions on faith, morality, or cosmic cycles. But this—this was something else entirely.

"The world does not exist as a single, unchanging truth," the Acharya continued. *"It is shaped by the eyes that see it. A river is not the same to the one who drinks from it as it is to the one who drowns in it. The universe does not reveal itself equally to all—it unfolds according to the awareness of the observer."*

Kartik felt his breath catch in his throat.

The observer changes reality.

This wasn't just philosophy—this was quantum mechanics.

His mind instantly recalled the double-slit experiment, one of the most baffling phenomena in physics. A particle, when unobserved, behaved as a wave, spreading out in probabilities, existing in multiple states at once. But the moment it was observed—the moment a conscious entity interacted with it—the wave collapsed into a single definite outcome.

It was as if reality itself was undecided, waiting for someone to look, to measure, to define it.

And now, here was a man speaking of the same concept—not in terms of electrons or probability waves, but as a fundamental truth of existence.

"The Upanishads say," the Acharya continued, *"'Drishti srishti vada'—that which is seen is created in the act of seeing. The moment*

you observe something, it is no longer what it was before. Consciousness is not separate from the world—it is woven into it."

Kartik swallowed hard.

If he had heard this yesterday, he would have dismissed it as poetic mysticism. But now, it sounded uncomfortably similar to the most cutting-edge discussions in physics.

Was consciousness truly an active participant in shaping reality?

If so, what was the true nature of existence before it was observed?

His rational mind resisted, but deep within, something unsettling was happening.

He had always believed that science uncovered truth by removing human bias, by stripping reality down to its core, measurable essence.

But what if reality itself depended on the observer?

What if the universe was not simply discovered—but, in some fundamental way, created by those who perceived it?

A shiver ran down his spine.

He had spent his life searching for objective truths. But now, he was beginning to wonder if such a thing even existed.

Consciousness: The Missing Equation?

The Acharya let the silence settle before speaking again. *"You believe the universe exists outside of you,"* he said, his voice calm but pointed. *"That it follows laws, independent of your presence, indifferent to your thoughts. But what if I told you that the mind is not merely an observer of reality—but its very architect?"*

Kartik stiffened.

He had indulged the parallels between ancient philosophy and physics so far, allowing curiosity to guide him. But this? This was dangerous territory.

The idea that consciousness shaped reality—that the mind was not just a passive receiver of information but an integral part of the universe itself—bordered on pseudoscience.

"We assume that the mind is contained within the body," the Acharya continued, *"that it is a byproduct of neurons, of chemical signals, of electrostatic charges firing in the brain. But if that were so, then why does perception alter reality? Why does the act of observation collapse possibility into form? Why does awareness hold the power to transform?"*

Kartik's jaw clenched.

He had heard this argument before—from speculative physicists, from consciousness researchers who toyed with the edges of quantum mechanics and philosophy. Some proposed that consciousness wasn't an emergent property of the brain at all—but a fundamental force, woven into the very fabric of existence.

But that theory had no measurable proof. No equations, no experimental validation—just a string of provocative questions.

"Thought is energy," the Acharya said, scanning the room. *"Every impulse in your mind is a wave, a vibration, a movement. And what is the universe but energy, moving in patterns we do not yet understand? If your thoughts are energy, and the universe is energy, can you truly say they are separate?"*

Kartik exhaled sharply.

The logic was deceptively simple. Too simple.

His mind raced through counterarguments. Correlation was not causation. The brain's electrical activity did not mean that thought created the universe, any more than a spark from a wire controlled a thunderstorm.

And yet...

His resistance faltered as he recalled something: The Penrose-Hameroff Orch-OR theory.

A controversial hypothesis, but one that suggested consciousness might originate at the quantum level—within the microtubules of neurons, where quantum effects could influence cognition. If true, it meant that the brain did not merely process reality—it participated in its formation at the smallest scale.

Could the Acharya be hinting at the same concept—only through a language far older than neuroscience?

Kartik shook his head. *No. I can't make that leap.*

Yet, the fact remained:

Physics had long ignored consciousness because it was immeasurable.

But what if it was missing from the equations not because it was irrelevant, but because science hadn't yet figured out how to include it?

His mind rebelled against the possibility, but for the first time, he didn't dismiss it outright.

Science, Myth, or Forgotten Knowledge?

The Acharya's voice remained steady, unwavering. *"What we call the past is not lost. Knowledge does not disappear—it only waits to be understood again."*

Kartik shifted in his seat, his mind caught between resistance and curiosity. Is that what was happening here? Was he merely seeing patterns where none existed, or had something been lost—not in destruction, but in translation?

The Acharya continued, "*The Vedic texts do not explain the world in formulas and equations. They speak in metaphor, in symbols, in concepts that require interpretation. But does that make them any less true?*"

He turned to the gathering. "*Let me ask you—if a child who has never seen the ocean is told that water is vast, deep, and unending, does his lack of experience make the statement false? Or does it only mean that, when he finally reaches the shore, he will understand what was described all along?*"

A murmur of quiet agreement passed through the audience. Kartik, however, was grappling with something deeper.

Metaphor.

That was the key, wasn't it? What if these ancient verses were not primitive explanations, but metaphorical descriptions of scientific truths—descriptions that modern science was only now catching up to?

The Acharya gave another example. "*The Rig Veda speaks of the universe originating from a single point, an infinitesimal seed of energy that expanded into everything. Is this not what modern physics now calls the Big Bang?*"

Kartik's breath hitched.

Yes. It was.

He had read theories proposing that early Hindu cosmology described a cyclic model of the universe—expanding, collapsing, and rebirthing in endless cycles, much like the oscillating universe model proposed by some physicists.

Coincidence?

"*And what about Indra's Net?*" the Acharya continued. "*The ancient teaching that the universe is like a great web, each point reflecting every other—infinitely interconnected. Is this not how modern physics describes quantum entanglement?"*

Kartik's fingers curled into his palms.

Entanglement.

Particles, separated by vast distances, influencing each other instantly—as if distance did not matter, as if everything was woven together in ways science still struggled to explain.

No. This was more than coincidence.

But was it possible that these concepts, these patterns, had been glimpsed long ago—not as mystical revelation, but as careful observation of the natural world?

Or was he merely bending meaning, searching for alignment where none truly existed?

His scientist's mind urged skepticism. Correlation is not causation. A poetic description of the cosmos did not mean the ancients had understood quantum mechanics.

But a quiet voice inside him pushed back. Then again, why did they describe it at all?

He had spent his life believing that science discovered truth.

But what if, in some cases, it was merely rediscovering what had already been known?

The Clash of Paradigms

Kartik felt the weight of contradiction pressing against him, two opposing forces battling for space in his mind. Everything he had trained for, every principle he had built his understanding upon, demanded that he remain skeptical.

And yet, skepticism was not the same as dismissal.

He had come here expecting to endure a philosophical discourse. Instead, he had been drawn into something far more unsettling—a mirror held up to his own field of study, reflecting ideas he could not ignore.

The Acharya's words did not contradict science outright; they echoed it, bent it, twisted it into something both familiar and foreign.

The Vedic descriptions of time—cyclical, flowing, nonlinear—were eerily similar to theories of relativity and quantum mechanics. The idea of Rta, a universal balance, sounded like an ancient articulation of the second law of thermodynamics, the interplay of order and entropy. The concept of Indra's Net bore a disturbing resemblance to the interconnected nature of quantum entanglement.

Coincidence? Or something more?

His rational mind resisted the urge to equate ancient philosophy with scientific rigor. Science was built on verification, on repeated experimentation, on mathematical proof. None of what the Acharya was saying could be tested, measured, or modeled.

And yet...

Kartik had spent his life searching for patterns in the universe. And here, before him, were patterns that had been seen and recorded thousands of years ago—without telescopes, without particle accelerators, without modern technology.

How?

The thought disturbed him because it left only two possibilities—either these were remarkable coincidences, poetic descriptions that happened to align with modern discoveries... or the ancients had perceived something beyond what conventional history credited them for.

He exhaled sharply. If he followed this path further, where would it lead?

Would he undermine his own scientific credibility? Would he begin seeing connections where none existed, grasping at illusions?

Or...

Would he discover something science had long overlooked?

The very thought made him uneasy.

Science had always been his anchor. But tonight, for the first time, he was beginning to wonder if it had also been his blindfold.

A New Question Takes Shape

The Pravachan came to an end, but the storm in Kartik's mind did not. Around him, the gathering began to dissolve—mourners whispering quiet farewells, a few staying back to speak with the Acharya. Yet, he remained seated, staring ahead but seeing nothing.

What if science is not discovering new truths, but rediscovering old ones?

The question had taken root somewhere deep within him, unsettling in its implications. He had spent his life believing that human progress was a steady climb—that knowledge moved in one direction, from ignorance to understanding, from myth to fact.

But what if he had been wrong?

What if the knowledge that modern science prided itself on uncovering had already been glimpsed—seen not through microscopes and equations, but through observation, intuition, and a different kind of inquiry?

He exhaled slowly, rubbing his temple. No. He wasn't ready to accept that. Not yet.

He was a scientist. If something was true, it had to be tested, proven, replicated. He could not—would not—accept knowledge simply because it was written in an ancient text or spoken with conviction.

And yet, he also could not ignore what he had heard.

The Acharya had not demanded faith. He had not spoken in absolutes or preached belief. He had only presented patterns—alignments between ideas that should not have intersected.

And now, those patterns refused to leave Kartik's mind.

A quiet voice inside him whispered, You came here to mourn, but you are leaving with a question you never expected to ask.

That, he realized, was the difference.

Skepticism had always been his shield. He had spent his life dismissing ideas that did not conform to his framework of reality.

But now?

Now, for the first time, he wasn't dismissing. He was questioning.

And questions, he knew, had a way of leading places he never intended to go.

As Kartik rose from his seat, stepping out into the cool night air, he knew one thing for certain.

This wasn't over.

Chapter 3: Seeds of Curiosity

The night air was thick with the scent of incense and damp earth as Kartik stepped outside the Pravachan hall. The gathering had dispersed, leaving behind only a few lingering conversations and the faint glow of oil lamps flickering against the temple walls. But inside his mind, the discourse had not ended.

He had come expecting to endure a sermon—something ritualistic, poetic, even comforting for those who needed faith to soften the weight of grief. Instead, he had encountered something far more unsettling.

The Acharya's words had not asked for belief. They had challenged understanding. They had spoken of time not as a straight path, but as a cycle, of reality not as an independent structure, but as something shaped by the observer. They had echoed concepts he had studied in physics, though they were never meant to be scientific.

And that was the part he could not ignore.

Kartik took a slow breath, his gaze drifting to the sky. The stars burned above, cold and indifferent, light traveling across space and time to meet his eyes. For the first time, he wondered—not as a physicist calculating distances, but as a thinker standing at the intersection of two seemingly different worlds.

What if knowledge was not always discovered, but sometimes remembered?

The thought should have made him uneasy. Instead, it planted something deeper. A question. A curiosity.

And curiosity, he knew, had a way of leading places he never intended to go.

A Mind in Turmoil

Kartik walked slowly through the dimly lit courtyard, his footsteps muffled by the cool earth beneath him. The sounds of the village had quieted, the evening settling into a hushed stillness. Yet, inside his mind, there was no such stillness.

The words of the Acharya clung to him, looping in fragments, forcing themselves back into his consciousness no matter how much he tried to dismiss them. *Reality is not fixed. Time does not move in a straight line. The observer shapes the observed.*

He exhaled sharply, running a hand through his hair. He had spent his entire life breaking the world down into measurable units, searching for patterns that could be tested, equations that could predict outcomes. That was science. That was truth.

But tonight, truth itself felt uncertain.

The Acharya had not presented proof, yet his words had felt disturbingly precise—as though he were describing modern physics, not ancient philosophy. *How was that possible?*

Kartik shook his head. *I'm seeing connections where there are none.* He had read about cognitive bias, the tendency to seek patterns even in randomness. Perhaps this was just that—his mind attempting to overlay scientific ideas onto poetic language.

And yet...

It didn't feel like coincidence.

His thoughts drifted back to the observer's paradox—the principle that in quantum mechanics, a particle's state is undefined until observed. *Reality is not independent of the one who sees it.* That was not spirituality. That was physics.

But if physics had only recently discovered this truth, how had ancient thinkers even conceptualized it?

He stopped walking, staring up at the vast night sky. The same sky that ancient sages had gazed upon, wondering about existence—just as he was now.

Were they asking the same questions?

Kartik inhaled deeply, trying to steady himself. The rational part of his mind urged him to let it go—to stick to what was verifiable, what was provable.

But another part of him, the part that had always sought knowledge beyond the obvious, whispered something else.

What if I'm not imagining these connections? What if I've been missing something all along?

For the first time, his skepticism was not enough to silence his curiosity.

A Chance Encounter

Kartik turned toward the temple gate, ready to leave, but something held him back. His steps were slow, his mind still entangled in questions he could not answer. The night air was cool, yet he felt a strange warmth—not comfort, but the lingering tension of an unresolved thought.

He had almost reached the exit when a voice cut through the silence behind him.

"You hesitate, Dr. Varun."

Kartik stopped. He hadn't expected to hear his name. He turned to find Acharya Vishwanath standing a few steps away, his expression unreadable.

Kartik straightened his posture instinctively. *"I was just... thinking,"* he said, though the words felt inadequate.

The Acharya nodded, his gaze calm but piercing. *"Thinking is good. But thinking alone can sometimes lead one in circles."* He took a slow step forward. *"Tell me, what troubles you?"*

Kartik hesitated. He wasn't sure if he wanted this conversation. He wasn't sure if he was ready to speak these thoughts out loud—because once spoken, they would demand an answer.

But something about the Acharya's presence made resistance feel unnecessary.

"I don't know if I'm troubled," Kartik admitted, crossing his arms. "But I do have questions."

The Acharya smiled. *"Questions are a beginning."* He gestured toward a stone bench under a large peepal tree at the temple courtyard. *"Shall we sit for a moment? Unless, of course, you fear that speaking about such things will lead you astray?"*

Kartik exhaled, shaking his head. *"I don't fear ideas, Acharya."*

"Good," the Acharya said, taking a seat. *"Then let us begin."*

Kartik hesitated only a moment before sitting down.

The night stretched quiet around them, but inside his mind, a storm was only beginning to form.

The Convergence of Thought

The night air was still, the distant flicker of oil lamps casting long shadows against the temple walls. Kartik sat with his elbows resting on his knees, his hands loosely clasped, his mind a battlefield of reason and doubt. Across from him, the Acharya sat in effortless composure, waiting—not pressing, not persuading, just allowing the silence to settle.

Finally, Kartik spoke.

"You say that ancient wisdom and science are not in conflict," he began, his voice measured, analytical. *"But I was taught that science progresses by questioning assumptions, by testing, by proving. Spirituality, on the other hand, asks for faith. How can they possibly align?"*

The Acharya smiled slightly, as if the question amused him. *"Do you believe that all truth must be proven before it is accepted?"*

Kartik hesitated. "In science, yes."

"And yet," the Acharya countered, *"many of the greatest scientific truths were known before they were proven."*

Kartik frowned, but before he could respond, the Acharya continued.

"The Earth was round before humans measured its curvature. Gravity existed before Newton formulated its laws. The atom existed before it was ever observed. Were these things false before they were proven?"

Kartik leaned back, exhaling. *"No, but the difference is that science eventually did prove them. Without proof, they were just ideas, not knowledge."*

The Acharya nodded. *"Then tell me, Dr. Varun—when physicists propose theories of dark matter, multiverses, or quantum entanglement, are they asking for faith?"*

Kartik stiffened. *"No, they are making hypotheses, awaiting validation."*

"Ah," the Acharya said, leaning forward slightly. *"And what if I told you that ancient knowledge operates in the same way—not*

demanding blind belief, but presenting hypotheses about reality? The difference is only in the method of inquiry."

Kartik blinked, caught off guard. *"You're saying that spirituality is just another form of theoretical science?"*

The Acharya chuckled. *"Not quite. Science seeks truth through observation and external validation. Ancient wisdom sought truth through internal realization and perception. The two paths are different, but that does not mean one is false."*

Kartik frowned. *"But internal realization is subjective. It cannot be measured."*

The Acharya inclined his head. *"And yet, reality itself may not be as objective as you think."*

Kartik's fingers curled slightly. The observer's paradox came rushing back to him. Quantum mechanics had already challenged the idea of an objective reality—had shown that observation itself could change outcomes.

He remained silent, and the Acharya continued.

"Science has built instruments to observe the outer world—telescopes to see the stars, microscopes to see the smallest particles. But the ancients believed that truth is not only outside—it is also within. Their instrument was the mind. Their laboratory was the self."

Kartik's skepticism flared again. *"But science is based on repeatability. If an experiment cannot be replicated, it is discarded. How can inner realization be considered valid when it cannot be tested?"*

The Acharya's gaze did not waver. *"Do you know what light is?"*

Kartik blinked. *"Of course. It's an electromagnetic wave, but also a particle. Its nature depends on how it is observed."*

The Acharya smiled. *"And yet, for thousands of years, people knew what light was without needing Maxwell's equations or quantum electrodynamics. They felt it, they saw its effects, they used it. They did not need proof to know it existed. Does that make their understanding invalid?"*

Kartik was silent.

"The mistake," the Acharya continued, *"is in thinking that science and ancient wisdom are two opposing forces. They are not. They are merely two languages attempting to describe the same reality. One does so through measurement, the other through perception."*

Kartik exhaled, rubbing his temple. He wanted to counter, to resist—but something in him hesitated. Because deep down, he knew the Acharya's words were not entirely wrong.

Hadn't he himself begun to suspect that modern science was only now rediscovering ideas that had been described long ago?

The Acharya studied him for a moment before speaking again, his voice softer now. *"You are not here to believe, Dr. Varun. You are here because you seek. And that is enough—for now."*

Kartik exhaled slowly, his mind a tangled web of possibilities.

For the first time, he was beginning to wonder if the division between science and wisdom had never truly existed—except in the way humans had chosen to frame them.

The Limits of Empiricism

Kartik leaned forward, his hands clasped, his mind balancing between skepticism and reluctant intrigue. *"Science is built on evidence,"* he said, his voice firmer now. *"It doesn't ask for faith. It asks for proof. If something cannot be tested, measured, or observed, then how can we call it knowledge?"*

The Acharya met his gaze with unwavering calm. *"And what if there are truths that exist beyond the reach of measurement?"*

Kartik exhaled sharply. *"Then they are just theories. Unverified speculation."*

The Acharya nodded slightly, as if expecting the response. *"Tell me, Dr. Varun—was the atom real before it was discovered? Was gravity untrue before Newton's equations?"*

Kartik hesitated. *"No, but those things were later proven."*

"Exactly." The Acharya leaned back. *"Science, too, has believed in things it could not yet prove. It predicted the existence of black holes long before they were observed. It proposed dark matter without direct evidence, based only on its effects. Does that mean those ideas were meaningless before proof arrived?"*

Kartik frowned. *"No, but those were still hypotheses grounded in observation. Ancient wisdom often deals in philosophy, not experiment."*

The Acharya raised an eyebrow. *"Does observation always require instruments? Must something be seen through a microscope or telescope to be real? The ancients observed in a different way. Their instruments were the mind, reflection, and experience. They did not measure the universe in numbers, but in patterns."*

Kartik opened his mouth to counter, but the Acharya continued.

"Tell me, Dr. Varun—how does science deal with paradox?"

Kartik hesitated, sensing a trap. *"By resolving contradictions through deeper understanding."*

The Acharya nodded. *"Then explain to me—how can light be both a particle and a wave? How can something exist in two states at once?"*

Kartik exhaled. "*That's quantum mechanics. It challenges classical logic, but the math supports it.*"

The Acharya smiled slightly. "*And yet, for centuries, people thought the world was purely deterministic. They believed everything could be predicted, like a machine. But science itself has had to evolve, to abandon its old assumptions when new discoveries emerged. Why, then, do you assume that your current knowledge is complete?*"

Kartik swallowed, his fingers tightening slightly.

The Acharya's voice softened. "*There was a time when science believed the universe was static, that time was absolute, that space was empty. Each of those ideas was later discarded. So tell me—if science is always revising itself, how can you be certain that what it rejects today will not be rediscovered tomorrow?*"

Kartik exhaled slowly.

He wanted to argue. He wanted to insist that science was different—that it progressed, while ancient knowledge remained fixed.

But even as he thought it, he realized it wasn't true.

Science had changed—again and again, it had rewritten its own understanding of reality.

And if knowledge could be lost and rediscovered in science... could the same be true for ancient wisdom?

The thought unsettled him.

For the first time, he wasn't sure whether he was defending science—or his own resistance to seeing beyond it.

Physics, Perception, and the Unseen

The Acharya's gaze remained steady as he spoke. "*Tell me, Dr. Varun—if a man who has never seen the ocean asks you to describe it, how would you explain?*"

Kartik frowned slightly, unsure where this was going. *"I'd tell him it's a vast body of water, deep and endless from the shore."*

The Acharya nodded. *"And if he says, 'I do not see this ocean. I do not feel its waves. Therefore, it does not exist'—what would you say?"*

Kartik exhaled sharply. *"I'd tell him that just because he hasn't seen it doesn't mean it isn't real."*

The Acharya's lips curved in a knowing smile. *"Ah. And yet, is that not the same reasoning science sometimes uses? If something cannot be measured, is it truly non-existent? Or is it simply beyond the reach of current instruments?"*

Kartik opened his mouth, then hesitated. The question was unsettlingly precise.

The Acharya continued, his voice calm. *"A few centuries ago, humans could not measure radio waves. Did that mean they did not exist? Before the first telescope, entire galaxies remained unseen—were they any less real?"*

Kartik's grip tightened on his knee. *"That's different. Those things were eventually proven."*

The Acharya inclined his head. *"Yes. But until they were proven, many doubted them. Just as today, there are things that exist beyond the boundaries of what science can measure. And yet, you assume that only what is measurable holds value."*

Kartik's mind was already searching for counterarguments, but a deeper part of him was unsettled—because this was precisely the problem modern physics was facing.

Science was built on observation and measurement, but what if reality itself was shaped by observation?

"The observer shapes the observed," the Acharya said, as if reading his thoughts. *"Your own field, quantum physics, has already shown*

that the act of measurement collapses probability into reality. The wave becomes a particle only when observed."

Kartik felt his pulse quicken. The double-slit experiment.

How had he not seen this connection before?

The Acharya leaned forward slightly. *"What does this mean, Dr. Varun? Does it mean that reality is waiting to be defined? That perception is not passive, but creative?"*

Kartik inhaled slowly. He had asked these same questions in physics discussions, in research papers, in speculative conversations with colleagues. But he had always kept them confined within the framework of science.

Here, the Acharya was asking the same questions, but through an entirely different lens.

"Science seeks to measure the universe," the Acharya continued. *"But what if the universe does not exist in a fixed state? What if, at its core, it is fluid, shaped by the very act of perceiving it?"*

Kartik swallowed hard.

If that were true, then reality itself was not independent. It was interactive.

And that—that changed everything.

A Scientist's Dilemma

Kartik let out a slow breath, his fingers interlocked as he stared at the temple courtyard, lost in thought. For the first time, he found himself at a loss for a definitive answer.

He had always believed that uncertainty was temporary—that science, with time, would illuminate every dark corner of the unknown. And yet, tonight, he was faced with questions that science had yet to even frame properly, let alone answer.

The Acharya observed him quietly, allowing the weight of the conversation to settle. *"You are troubled."*

Kartik shook his head. *"Not troubled. Just... processing."*

The Acharya smiled faintly. *"Processing is good. But tell me, Dr. Varun, why does uncertainty make you uneasy?"*

Kartik exhaled. *"Because I was taught that if we don't know something, we just need to ask the right questions, run the right experiments, and eventually, we'll find an answer."*

The Acharya nodded. *"And what if the answer is not found in equations, but in something beyond measurement?"*

Kartik hesitated, his rational mind immediately resisting. *"Then it remains an idea, not knowledge."*

The Acharya's expression did not change. *"Ah. And yet, didn't quantum mechanics remain only an idea until it was observed? Didn't black holes exist only in mathematical theories before they were proven?"*

Kartik's jaw clenched. He knew the point the Acharya was making. Science itself had often dismissed concepts until it was forced to accept them.

He had spent his life believing that knowledge followed a linear path—that each discovery built upon the next, bringing humanity closer to understanding reality.

But what if science was not discovering, but rediscovering?

The Acharya's voice was steady. *"Dr. Varun, have you considered that knowledge does not always progress forward, but sometimes in cycles—just as time, just as existence?"*

Kartik looked at him, realizing the irony. Hadn't he just argued against linear time? Hadn't he entertained the idea that time itself may be cyclical?

If that was true, then why not knowledge itself?

His own words came back to him: *What if science is not discovering new truths, but rediscovering old ones?*

He exhaled sharply. He did not have an answer.

For the first time in his career, he had encountered a question that science had yet to quantify—and he wasn't sure if it ever could.

A Door Left Open

The temple courtyard was quiet, save for the occasional rustling of leaves in the evening breeze. The oil lamps flickered, casting elongated shadows on the stone floor. The conversation had reached its end, but the debate within Kartik had not.

He looked at the Acharya, expecting a final argument, a concluding remark that would push him toward acceptance or dismissal. Instead, the Acharya simply watched him, his expression calm, almost amused, as if he knew Kartik's own thoughts better than he did.

"You seek proof," the Acharya finally said, his voice steady. *"But what if proof comes only after belief?"*

Kartik inhaled sharply. The question unsettled him.

Science did not work that way. Belief was a bias—an obstacle to objective observation. Scientists did not start with belief; they started with skepticism and let the evidence guide them.

And yet...

Wasn't the history of discovery filled with moments where scientists had to entertain the possibility of an idea before they had the tools to prove it?

Didn't Einstein propose relativity before space-time curvature was confirmed?

Didn't physicists believe in the Higgs boson before they built the collider to detect it?

Hadn't science, too, required faith in an idea before the proof was visible?

The realization struck deeper than he wanted to admit.

Kartik exhaled. *"That's a dangerous way of thinking."*

The Acharya smiled. *"For those who fear being wrong, perhaps."*

Kartik looked away for a moment, then back at the Acharya. *"I don't fear being wrong."*

"No," the Acharya agreed. *"You fear what happens if you're right."*

The words landed heavier than Kartik expected.

Because he knew what it meant. If there was truth in what the Acharya had said, then reality was not just what science could measure. It was more.

And that would mean that his understanding of knowledge itself was incomplete.

The Acharya did not press further. He simply stood, adjusting his robes. "The night is deep. You have much to think about."

Kartik nodded slowly. *"I do."*

The Acharya turned toward the temple steps but paused. *"Dr. Varun,"* he said over his shoulder, *"understanding does not begin with proof. It begins with the willingness to see differently."*

And then, he walked away.

Kartik remained seated for a long time.

He should have felt relieved that the discussion was over. Instead, he felt unsettled—because now, he knew he could not ignore this any longer.

He had come seeking answers, but all he had found were better questions.

And questions, he knew, had a way of demanding pursuit.

As he rose to leave, the night stretched before him—wide, open, and waiting.

Part II — Immersion Into The Ancient Lens

Where skepticism meets symbol, and equations meet metaphors

Chapter 4: Unveiling the Scriptures

The glow of Kartik's laptop screen flickered against the darkened room, its cold blue light illuminating the stacks of books and scattered notes across his desk. He had spent years surrounded by scientific journals, research papers, and equations that neatly fit within the boundaries of logic and proof. But tonight, his workspace looked different.

In front of him lay a series of ancient texts, their pages delicate, their language dense with metaphor and meaning that eluded straightforward analysis. The Rig Veda. The Upanishads. The Bhagavad Gita. He had never imagined he would be reading them—not with intent, not like this.

Yet here he was.

His conversation with the Acharya had left him restless. He wasn't here to find faith—he was here to test an idea. Were these scriptures merely spiritual doctrines, or did they contain something more—a forgotten layer of knowledge, buried beneath metaphor and tradition?

He had always dismissed religious texts as unstructured, imprecise—the opposite of scientific thought. But as he traced his fingers over the verses, reading their rhythmic, poetic expressions of time, space, and existence, he found himself asking:

Were these early attempts at explaining the universe?

Had the ancients, in their own way, described what physics was only now beginning to uncover?

The thought made him uneasy. But he couldn't walk away.

Not yet.

He had spent his life searching for truth. And now, he had to know if it had been hidden in these words all along.

A Scholar in Unknown Territory

Kartik leaned back in his chair, rubbing his temples as he stared at the open book before him. The Rig Veda lay in front of him, its pages filled with verses that spoke in riddles—hymns about creation, cosmic order, and the fabric of reality, yet written in a language that resisted direct interpretation.

This was not like reading a scientific paper.

There were no equations, no structured hypotheses, no controlled experiments that led to clear conclusions. Instead, there were metaphors wrapped in poetry, allegories that hinted at something beyond their literal meaning.

He sighed and closed his eyes for a moment. The scientist in him wanted clarity—direct statements that either confirmed or rejected an idea. But these texts did not operate that way. They spoke in layers, each line containing multiple meanings, requiring interpretation rather than calculation.

He flipped back to a passage he had marked earlier, one that had caught his attention.

"Then was neither non-existence nor existence: There was no realm of air, no sky beyond it. What covered it? Where was it? In whose keeping? Was there then cosmic waters, in depths unfathomed?"

Kartik exhaled sharply. A verse from the Nasadiya Sukta—the Hymn of Creation.

At first glance, it seemed poetic, even vague. But when he read it again, he realized it was describing a time before the universe existed—before space, before form.

And wasn't that exactly what modern cosmology sought to understand?

He grabbed his notebook and scribbled a comparison.

- Vedic Text: *"Neither non-existence nor existence"* → A description of a state before the universe began?
- Modern Physics: The pre-Big Bang singularity—a state where conventional physics breaks down, where even time has no meaning.

Kartik frowned. Could this be coincidence? Or was it possible that these ancient philosophers had intuitively grasped something that modern physics was still struggling to define?

He shook his head. He needed more evidence.

Frustration crept in as he flipped through more pages. Understanding these texts was like deciphering an ancient code. The problem wasn't just the language—it was the thought process behind them.

Science followed a linear path—observation, hypothesis, testing, conclusion. These texts, however, did not reveal knowledge in a straight line.

Instead, they described reality through experience, intuition, and metaphor.

Kartik ran a hand through his hair. He was not used to this kind of thinking.

For the first time in his career, he felt like a student again—not of physics, but of something far older.

And despite his frustration, despite his resistance, he knew he couldn't stop now.

The Weight of Interpretation

Kartik exhaled sharply, setting his pen down beside his notebook. His desk was a mess of open texts, hastily scribbled notes, and bookmarks jutting out of pages like fragmented thoughts. The deeper he read, the more elusive the answers became.

This was not how knowledge was supposed to work.

In physics, a formula explained a phenomenon, a proof validated a theory. But here, truth was hidden beneath layers of metaphor, buried in verses that refused to speak plainly.

He flipped back to a passage he had read earlier from the Upanishads:

"That which is the smallest of the small and the greatest of the great, That which is hidden in the secret place of the heart, That which is unchanging while all things change— That is the eternal reality."

Kartik frowned, his fingers tightening on the page. What did this mean?

Was this describing the concept of the atom, the indivisible building block of matter? Or was it speaking of something more abstract—consciousness, existence, the self?

It was impossible to tell.

His frustration deepened. If the ancients had glimpsed truths about reality, why didn't they just say it directly?

Why wrap knowledge in riddles? Why encode ideas instead of explaining them?

Kartik stood and walked to the window, staring out at the dark sky.

Was this spiritual mysticism, or had these thinkers simply operated with a different kind of logic—one that modern science was unequipped to understand?

He had always believed that truth was best communicated through precision—through mathematics, through measurable proof.

But what if there were things that could not be described in equations?

What if the nature of certain truths required a language beyond structured logic?

He ran a hand through his hair, sighing. He wasn't ready to accept that yet.

But one thing was clear: if these texts held any real knowledge, he would need more than logic to understand them.

He needed a different way of seeing.

Seeking Guidance

Kartik knew when he had reached the limits of his own expertise. No matter how many times he reread the texts, he lacked the tools to decode them properly. The language, the layered meanings, the references woven into poetic verses—it was an intellectual maze.

So he did what any scientist would do when confronted with an unfamiliar field. He sought experts.

After a few calls and emails, he secured a meeting with Dr. Madhav Shastri, a well-regarded scholar of Sanskrit and ancient Indian philosophy. A man of both academia and tradition, Dr. Shastri had spent decades studying Vedic literature, its historical context, and its evolving interpretations.

Their meeting took place in the professor's study, an old, dimly lit room filled with shelves of ancient manuscripts, palm-leaf texts, and modern research papers.

Dr. Shastri, a man in his sixties with piercing eyes behind thick glasses, leaned forward, studying Kartik with interest. *"So, a physicist seeks answers in the Vedas. That is rare."*

Kartik offered a faint smile. *"I wouldn't say I'm seeking answers. I'm... testing an idea."*

The professor chuckled. "That's how it always begins."

Kartik hesitated before placing an open copy of the Rig Veda on the table. "I'm struggling to understand how knowledge was structured in these texts. There are references that seem almost scientific, yet they are hidden in poetry and metaphor. Why? Why not just state the truth plainly?"

Dr. Shastri leaned back, folding his arms. *"Because, Dr. Varun, Vedic knowledge was not meant to be read—it was meant to be realized."*

Kartik frowned. *"What do you mean?"*

The professor gestured toward the book. *"Unlike modern texts, which explain facts directly, the Vedas were transmitted through oral tradition for centuries. They were not just books of knowledge—they were living philosophies, designed to be memorized, recited, and understood gradually, layer by layer."*

Kartik listened intently.

"The language of the Vedas is deliberately layered. A verse has multiple meanings—one for the ordinary listener, one for the philosopher, one for the seeker of truth. The same words may describe a fire ritual on the surface but hint at cosmic principles when viewed differently."

Kartik's fingers tightened on the book. This was not just about ancient literature—it was an entirely different way of encoding knowledge.

Dr. Shastri continued. *"The rishis did not write explanations because they believed that true understanding comes not from reading, but from experiencing and contemplating. Knowledge was not handed out—it had to be earned."*

Kartik exhaled. *"So the meaning is there... but hidden."*

Dr. Shastri smiled. "Not hidden. Just waiting for the right perspective."

Kartik glanced back at the verses in front of him. Perhaps he had been looking at these texts the wrong way all along.

Maybe they were not meant to be read like equations, but unraveled like a code.

And if that was true—then he had only just begun.

The Concept of Rta: Cosmic Order or Natural Law?

Kartik sat at his desk, flipping through the pages of a translated commentary on the Rig Veda. His discussions with Dr. Shastri had shifted his approach—he was no longer searching for direct statements that mirrored scientific laws. That was a mistake he had made earlier, trying to fit ancient words into modern frameworks. Instead, he was now looking for patterns—underlying principles that might suggest a deeper, structured understanding of reality.

And that was when he found it.

Rta.

The word kept surfacing in different contexts, woven through hymns about creation, celestial motion, natural cycles, morality, even truth itself. The texts spoke of Rta as the fundamental order of existence—an invisible framework that maintained balance across the physical, ethical, and cosmic realms.

Kartik leaned forward, his fingers tracing the word on the page. Order.

His mind instinctively drew a parallel—physics, at its core, was also a search for order. Scientists had spent centuries uncovering the hidden laws that governed everything from the behavior of subatomic particles to the expansion of the universe.

Was this what the ancients had glimpsed—a recognition that the universe was not chaotic, but structured?

He flipped to another verse.

"The sun rises in Rta, the stars shine in Rta, the waters flow in Rta, the seasons turn in Rta."

It read like poetry. But when stripped of its literary beauty, was this not a fundamental description of natural laws?

- The sun rises → Celestial mechanics, governed by gravity.
- The stars shine → Nuclear fusion, the energy balance sustaining the cosmos.
- The waters flow → Fluid dynamics, the hydrological cycle.
- The seasons turn → Earth's axial tilt, orbital mechanics.

Kartik's pulse quickened. Had the ancients, in their own way, recognized the structured nature of reality—long before equations described it?

He jotted down a thought in his notebook:

- Modern Science → Describes laws with precise mathematical formulations.
- Vedic Thought → Encodes the same principles as an

observed and experienced order in nature.

Two different approaches. Same reality.

He flipped through more texts, looking for additional references. The descriptions of Rta consistently pointed to balance—not just in the sense of moral order, but as a force that prevented the universe from collapsing into disorder.

A realization struck him.

Entropy.

In thermodynamics, entropy measured the degree of disorder in a system. The universe naturally moved toward higher entropy, yet order persisted within localized systems—planetary orbits, atomic structures, even biological life.

Had Rta been an early recognition of this delicate equilibrium—the interplay between chaos and order?

His skepticism flared again. He wasn't ready to call this science—not yet. Science required rigorous testing, repeatability, quantifiable models.

But he could no longer dismiss Rta as just spiritual abstraction, either.

These texts were not presenting hypotheses in the way science did—but what if they had been attempting to describe the same fundamental truths, just through a different mode of expression?

Kartik exhaled, staring at the page in front of him.

If that were the case, then the boundary between science and philosophy was thinner than he had ever imagined.

And if Rta was not just poetic mysticism but a coded articulation of universal balance—then what else had these texts understood before science could measure it?

Hidden Patterns in Vedic Verses

Kartik flipped through the pages, his pencil tapping absently against the margin of his notebook. The more he read, the more uneasy he felt. There were patterns here—hidden, woven beneath the surface of poetic verses.

And some of those patterns hinted at a structure beyond metaphor.

One passage from the Yajur Veda caught his attention:

"That which moves and that which moves not, That which is far and that which is near, That which is within and that which is beyond— That alone is the eternal truth."

Kartik's grip on the book tightened. He read it again. *That which moves and that which moves not.*

This wasn't just poetic paradox. It suggested something fundamental: motion and stillness were not absolute properties, but relative states dependent on the medium in which they existed.

His mind wandered to structured energy dynamics. In an energy-based model, motion is not an inherent property of an object, but a result of energy interactions. A system could appear still yet be fluctuating at an unseen level. A quantum system, when stripped of its probabilistic interpretations, could be a structured field where motion and stillness coexist in an equilibrium that maintains reality.

And then there was the next part—*far and near, within and beyond.*

This was not quantum entanglement in the mainstream sense, where particle states seem "linked" across distances in a way that defies locality. Instead, it spoke of a structured medium, where all interactions occurred within the same field, making distance irrelevant. Much like waves in an ocean—one could measure separation, but the medium remained continuous.

He turned the page, scanning for more references, and found another verse from the Upanishads:

"As is the atom, so is the universe. As is the human body, so is the cosmic body. As is the microcosm, so is the macrocosm."

Kartik inhaled sharply.

This was a profound statement—not a mystical claim, but a direct assertion of self-similarity in nature. In physics, this was known as scale invariance—the principle that the same structures repeated at different scales. Spiral galaxies mirrored hurricanes. The flow of electrons around a nucleus followed the same energetic tendencies seen in planetary motion.

But more than that—this hinted at structured vacuum energy. If the universe was an interconnected field of structured energy, then the patterns seen at the smallest scale would be reflected in the largest. Matter, in this view, was not an independent entity but a manifestation of structured energy configurations—stable yet dynamic.

Kartik rubbed his temple. Were these merely poetic observations, or had the ancients glimpsed an underlying truth?

The unsettling part was that these texts were not attempting to *prove* anything. They did not present theories, arguments, or hypotheses to be debated.

They simply stated reality as it was—without needing to justify it.

Kartik closed the book and leaned back in his chair. His rational mind resisted, but a deeper part of him sensed something undeniable.

Had the ancients understood energy as the true foundation of reality—before science had the tools to measure it?

And if so...

What else had they known?

Science and Spirituality: A Parallel Evolution?

Kartik stared at the open book before him, but his eyes were no longer focused on the words. His mind was caught in a loop of contradictions.

For years, he had believed that science was the only true path to knowledge—built on experimentation, measurement, and repeatability. Anything outside of that was, at best, philosophical speculation and, at worst, superstition.

And yet, here he was, confronted by something he had not expected.

Ancient thinkers had described reality in ways that resonated with fundamental scientific principles, not through mathematical models, but through direct observation and experience. They had spoken of order in the universe, structured energy interactions, and the cyclical nature of existence—not as abstract beliefs, but as a lived understanding.

He flipped back to the verse on Rta, the cosmic order that maintained balance in all things. Had they intuited a governing principle of structured energy before physics had even developed the tools to measure such interactions?

Modern science sought truth through experiments, equations, and empirical models.

Ancient wisdom sought truth through direct perception, experience, and introspection.

Two different methodologies. One shared pursuit.

Kartik ran his fingers through his hair. Was modern science truly "discovering" these concepts, or was it merely finding ways to articulate what had already been understood in a different form?

For the first time, he questioned something deeper.

Had human knowledge always been cyclical, rather than linear?

History had already shown that civilizations lost and rediscovered knowledge over time. The heliocentric model of the solar system, once suggested in early Indian and Greek texts, had been forgotten for centuries before Copernicus revived it. The medical understanding of consciousness and mind-body connections, central to ancient Indian texts, was now being revisited through neuroscience and quantum cognition.

What if this was another cycle?

What if modern physics, instead of leading humanity into new frontiers of thought, was merely retracing a forgotten path?

He exhaled sharply.

This wasn't about abandoning science in favor of ancient wisdom. He wasn't about to discard the rigor of empirical inquiry.

But he could no longer deny that science and philosophy were not opposing forces—they were evolving in parallel.

Each offered a different lens, but both were searching for the same truth.

And perhaps, to truly understand the universe, neither could be ignored.

Between Skepticism and Discovery

Kartik closed the book, exhaling slowly. His fingers lingered on the worn edges of the pages, as if the weight of the words had somehow transferred to his hands. He had expected to disprove, to dismantle, to rationalize these texts as poetic musings of a less advanced time.

Instead, he found himself at an uncomfortable crossroads.

The more he read, the more his resistance weakened.

He had spent his life trusting the structure of science—hypothesis, experiment, validation. If something could not be measured, it was not real. That was the foundation of modern knowledge.

And yet, these verses did not demand belief. They simply stated reality, with a quiet certainty that unsettled him.

He picked up his notebook and glanced at the list of parallels he had jotted down between Vedic concepts and physics:

- **Rta** – Not just cosmic order, but a structured field of equilibrium, an early articulation of the energy medium that science was still struggling to define.
- **Interconnectedness** – Not quantum entanglement as popularly misinterpreted, but an understanding that all interactions occurred within the same structured medium, where distance and locality were perceptual constructs.
- **Microcosm and Macrocosm** – Not vague philosophy, but a direct recognition of scale-invariant structures, self-similar energy configurations across different magnitudes.

His logical mind still resisted. Was this knowledge, or was he merely seeing patterns where none existed?

The texts lacked equations, lacked structured methodologies, lacked falsifiability—the very pillars of modern science.

But at the same time, science itself had evolved by repeatedly discarding its own flawed assumptions.

- Newton's absolute space and time had been overturned by Einstein.
- The certainty of classical determinism had collapsed under quantum mechanics.
- The idea of an empty vacuum had been replaced with the reality of structured vacuum energy.

What if modern science's framework itself was incomplete?

His fingers tightened around the notebook. He was not ready to accept that these texts were equivalent to scientific models—but he also could not ignore what he had seen.

He was standing at an intellectual threshold.

Either he dismissed these insights as poetic wisdom with no measurable proof, closing this chapter in his mind.

Or...

He continued seeking, challenging his own understanding, and exploring the possibility that knowledge had always existed in multiple forms.

Kartik leaned back in his chair.

For a man who had spent his life searching for answers, there was only one choice.

He had to keep going.

A New Direction

Kartik took a deep breath, staring at the pages spread before him. He was done trying to force these texts into the rigid framework of modern science. That approach had only led to frustration.

Instead, he would treat them as what they truly were—a different system of knowledge, encoded in metaphor, layered with meaning, meant to be unraveled rather than read at face value.

This was not religious doctrine.

This was an intellectual system that preserved insights in a way that required interpretation, not blind acceptance.

His mistake had been assuming that truth must always be structured in numbers and formulas. But history had already proven that different civilizations had recorded knowledge in different ways:

- The Greeks used geometry and axioms.
- The Chinese developed symbolic correlations in natural cycles.
- The Indians relied on linguistic structures, mantras, and layered interpretation.

Perhaps the Vedic texts were not offering answers outright, but encoding a way to reach them.

He flipped back to a verse from the Upanishads:

"Neti, Neti—Not this, not that."

It had always been translated as a way of negation—to remove assumptions, to discard falsehood, to arrive at the truth by eliminating the untrue.

Kartik tapped his pen against his notebook. Is this not what science does?

At its core, scientific progress was a process of negation—discarding flawed models, refining theories, challenging assumptions.

Perhaps, instead of opposing science, these texts were describing the same process—just in a different way.

He exhaled. This was a breakthrough.

From now on, he wouldn't reject these ideas simply because they were not structured in the way he had been trained to understand them.

He would decode them, extract the essence of their insights, and see where they aligned with the structured reality science sought to define.

This wasn't about faith.

This was about a different way of thinking.

Kartik closed his notebook with renewed determination. He was no longer just a skeptic. He was a seeker.

Chapter 5: Bridging Two Worlds

Kartik stood in his lab, staring at the stream of data on his screen. Numbers, waveforms, energy distributions—all neatly plotted, all undeniably real. And yet, they pointed to something he wasn't ready to accept.

He exhaled, rubbing his temples. This wasn't supposed to happen.

He had begun this journey out of curiosity, driven by the intellectual challenge of decoding ancient wisdom, of testing whether there was any merit in these Vedic descriptions of reality. But he had never expected to find results that aligned so precisely with what the texts described.

He glanced at his notebook, where he had scribbled his latest observations alongside a passage from the Rig Veda. He hadn't forced the connection—it was just there.

The scientist in him resisted. Correlation is not causation.

And yet, the data did not lie.

For the first time in his career, he was standing between two worlds—one built on equations, the other on intuition.

And the gap between them was far smaller than he had ever imagined.

From Interpretation to Experimentation

Kartik paced across his lab, his mind racing faster than his steps. For weeks, he had been buried in the Rig Veda, the Upanishads, and Sanskrit commentaries, deciphering patterns, extracting meanings, and mapping them onto scientific principles.

But something was missing.

Interpretation alone was not enough.

Vedic texts described reality through experience, but science demanded validation. If there was truly a bridge between ancient wisdom and modern physics, he needed to test it.

He sat at his desk and flipped through his notes. Among the pages of scribbled equations and translated Sanskrit verses, he had written down a fundamental realization:

Ancient texts describe universal order not as rigid laws, but as emergent principles—self-regulating, self-sustaining.

This was an entirely different way of thinking.

Science sought absolute measurements and constants, while Vedic thought described dynamic balance and interconnection.

Could these two perspectives be reconciled?

Kartik knew he had to start somewhere. He needed a hypothesis—something tangible, measurable, something that could withstand the rigor of experimentation.

His gaze fell on his latest notes about Rta—the principle of universal balance.

He had been viewing Rta as a philosophical abstraction, but what if it wasn't? What if it was describing an underlying energy structure—an equilibrium that governed the universe, much like structured vacuum energy?

He grabbed a fresh page and started writing.

Hypothesis: If Vedic insights describe structured energy interactions, there should be observable patterns that align with physical laws—ones that modern science has yet to fully quantify.

He leaned back in his chair.

This was it.

The first step.

Not proving the Vedas as "scientific" documents, not forcing connections, but using them as a framework for asking new questions.

Science had always evolved by challenging assumptions. Maybe it was time to challenge the assumption that knowledge had only one language.

Designing the First Experiment

Kartik stared at the blank page in his notebook, tapping his pen against the desk. Where to begin?

He had formulated a hypothesis—that Vedic insights, particularly Rta, described an underlying energy structure maintaining balance in the universe. But translating that into an experiment was a different challenge altogether.

Vedic texts did not describe reality in numbers or equations. They spoke in principles—balance, harmony, cyclic motion, the unseen forces shaping the seen world. But what if these ideas hinted at measurable physical interactions?

He flipped through his notes, stopping at a passage from the Rig Veda:

"The order of the universe flows through all things, unseen yet present, shaping movement and stillness alike."

He had originally dismissed it as metaphor. But now, it reminded him of something else—the structured nature of energy fields.

In modern physics, space wasn't empty; it was filled with energy fluctuations, fields that determined how matter and forces interacted. Could the ancients have recognized this structured medium, not through equations, but through observation and philosophy?

He set his pen down and started listing parameters:

- **Concept to test:** The existence of structured energy balance in a controlled system.
- **Scientific basis:** The idea that space is not truly empty, but an active medium.
- **Vedic correlation:** Rta as the principle governing this unseen yet ever-present energy.

Now came the hardest part—testing it.

Building a Controlled System

Kartik needed to design an experiment that could detect structured balance in energy fluctuations, rather than just random noise. He considered modern experimental setups that measured vacuum energy or spontaneous energy patterns.

One approach stood out—interference experiments.

- If Rta described a fundamental medium of balance, its influence should appear in wave interference patterns.
- He could use laser interferometry, tracking small-scale energy fluctuations in a controlled system.

- If there were persistent, structured variations, it could suggest an underlying equilibrium field—something that modern physics hadn't fully accounted for.

He sketched the setup in his notebook—a laser beam split into two paths, recombining to form an interference pattern. Small shifts in energy structure could alter the pattern in measurable ways.

He exhaled. This was not about proving the Vedas scientifically—it was about seeing if their insights aligned with measurable reality.

Now, he had a testable framework.

The real question was—what would he find?

The First Anomalies

Kartik leaned over his workstation, eyes fixed on the screen displaying the live feed of his laser interference experiment. The setup had been running for hours, logging minute fluctuations in energy patterns. The experiment was supposed to be a baseline test—a controlled setup designed to detect subtle shifts in structured energy.

But what he was seeing made no sense.

He had expected random noise, small fluctuations that could be accounted for by external factors—thermal drift, instrument sensitivity, even minor vibrations. Instead, the data showed something else.

A pattern.

It wasn't a perfect repetition, but there was an unmistakable periodicity in the shifts—something was self-correcting, stabilizing within the interference fringes.

He rubbed his temples. This wasn't supposed to happen.

He cross-checked the logs, recalibrated the instruments, re-ran the system with controlled variations. Still, the pattern persisted.

He pulled out his notebook and flipped back to his notes on Rta.

"The order of the universe flows through all things, unseen yet present, shaping movement and stillness alike."

Was this the balance he had read about—an unseen self-regulating structure embedded in energy itself?

His rational mind rejected the idea immediately. There had to be another explanation.

Coincidence?

Instrumentation error?

An overlooked external factor?

He checked everything again. No anomalies in the equipment. No environmental interferences. Yet, the fluctuations in the interference pattern continued to self-correct—as if the system itself was governed by an underlying equilibrium.

Had he stumbled onto something real?

Or was he seeing what he wanted to see?

Kartik pushed back from the desk, exhaling sharply. If this was genuine, it could mean that energy did not just fluctuate randomly—it followed an intrinsic order, a principle of balance that modern physics had yet to quantify.

But if he was wrong, if this was just an artifact of his own experimental setup...

He needed more tests. More data.

For now, all he had was a pattern—one that aligned eerily with Vedic descriptions of universal balance.

And he wasn't sure if that excited him or terrified him.

Skepticism Among Peers

Kartik stood at the front of the conference room, arms crossed, as his laptop projected the interference patterns onto the large screen. The room was silent.

He had spent the past hour walking his colleagues through his experiment—the setup, the controls, the results. He had deliberately avoided mentioning Vedic philosophy or any reference to ancient texts. This was pure data, nothing more.

Finally, one of the senior researchers, Dr. Mehta, broke the silence.

"This is... interesting," he said, leaning forward. *"But what exactly are you proposing here, Kartik? That space itself has some kind of self-regulating order?"*

Kartik exhaled. *"Not exactly. The data suggests that energy fluctuations within the system don't behave as purely random noise. There's a self-stabilizing structure at play. That's what I'm trying to understand."*

Dr. Mehta frowned. *"And what theoretical framework are you basing this on?"*

Kartik hesitated. This was the moment.

He could stick to conventional terminology—talk about energy fluctuations, hidden variables, or structured fields in vacuum dynamics. Or...

He could tell them the truth.

He took a breath.

"There are descriptions in ancient Indian texts," he began carefully, *"of a fundamental principle called Rta—a concept that refers to an intrinsic order governing reality. It speaks of balance, of a structured medium through which all things interact."*

A shift in the room. Some eyebrows raised. Some faces hardened.

"I'm not saying this proves anything," Kartik continued quickly. *"But the alignment between these insights and what we're seeing in the data is... unsettling. It raises a question—could these texts have described fundamental principles that science is only now beginning to quantify?"*

The skepticism was instant.

Dr. Verma, a physicist Kartik had always respected, scoffed. *"Come on, Kartik. You're telling me an ancient philosophy somehow predicted structured energy interactions? Without equations? Without instruments?"*

Another researcher, Dr. Rao, was more measured. *"It's not completely unreasonable,"* he said slowly. *"Historically, there have been cases where ancient knowledge turned out to be valuable once we had the tools to measure it. But..."* He hesitated. *"Aligning this with Vedic philosophy is a stretch."*

Kartik sighed inwardly. He had expected this.

Science was built on skepticism, and for good reason. But why was it so difficult to accept that knowledge might have been preserved in a different form?

Dr. Mehta folded his arms. *"You have something interesting here, Kartik. But if you want anyone to take it seriously, you need to explain it in terms of established physics. The moment you bring in philosophy, you'll lose credibility."*

Kartik felt a sharp tension in his chest.

He understood their caution. But was scientific inquiry supposed to have boundaries?

The meeting ended with no clear conclusions.

Some were intrigued but hesitant to associate this with ancient knowledge. Others dismissed it outright.

Kartik packed up his laptop and left the room, knowing one thing for certain—

If he wanted answers, he would have to find them himself.

Refining the Hypothesis

Kartik sat in his office, his notes spread across the desk, a cup of untouched coffee growing cold beside him. The skepticism from his peers hadn't discouraged him—it had clarified his next step.

His approach had been flawed.

He had presented data but lacked a framework that made it scientifically digestible. He had expected his colleagues to see the connections, but science didn't work that way—it demanded structured interpretation, not philosophical speculation.

If he wanted this to be taken seriously, he needed to speak the language of science while preserving the depth of Vedic thought.

A Multi-Disciplinary Approach

Kartik realized he could no longer work in isolation. His research required insights from multiple fields:

Physics – To refine the structured energy model and eliminate conventional explanations for the anomalies.

Mathematics – To develop a quantifiable framework that could model the observed self-regulating energy fluctuations.

Sanskrit & Vedic Studies – To ensure he wasn't misinterpreting the ancient descriptions, but instead extracting their essence correctly.

With this in mind, he began reaching out.

Consulting Theoretical Physicists

He met with **Dr. Ananya Roy**, a theoretical physicist specializing in vacuum energy interactions. He carefully presented his data without mentioning Vedic philosophy.

She studied his results for a long moment before saying, *"This is... unusual. You're suggesting that the fluctuations aren't entirely stochastic?"*

"Right," Kartik nodded. *"There's a self-correcting behavior—almost like an equilibrium within the system."*

Dr. Roy tapped her pen against the desk. *"That would mean there's an underlying structure regulating the fluctuations. That's not how we currently model vacuum interactions, but..."* She hesitated. *"It's an interesting anomaly. I'd be curious to see if it holds up under different conditions."*

It was a start.

Seeking Mathematical Rigor

To build a solid model, Kartik reached out to **Dr. Narayan Iyer**, a mathematician working on nonlinear dynamical systems.

"I need to develop a framework to quantify energy fluctuations that exhibit self-regulation," Kartik explained. "Something that can map equilibrium patterns over time."

Dr. Iyer leaned back in his chair. *"You're talking about a feedback-driven system. Have you considered applying control theory?"*

Kartik frowned. He hadn't thought of that. Control theory was used in engineering and cybernetics, but could it explain fundamental energy interactions?

"I can help you model the feedback loops," Dr. Iyer continued, *"but you need to define what's governing this balance."*

That was the missing piece.

Revisiting the Vedic Descriptions

Kartik returned to his Sanskrit sources, consulting **Dr. Shastri** once again.

"You're telling me that Rta is not just cosmic law but a self-regulating force?" Kartik asked.

Dr. Shastri nodded. *"Yes, the Vedic texts describe Rta as something dynamic—not a fixed law, but a principle that ensures harmony by constant adaptation. It is the balance that sustains itself."*

Kartik's breath caught. Feedback loops.

That was exactly what he was observing in the experiment—energy fluctuations stabilizing themselves within a system.

Could Rta have been an early conceptualization of self-organizing energy dynamics?

A More Structured Hypothesis

With these insights, Kartik refined his original hypothesis:

- The observed energy stabilization patterns suggest an intrinsic feedback mechanism within structured energy interactions.
- This aligns with the Vedic principle of Rta, which describes self-regulating order within existence.
- If validated, this could provide a missing link in our

understanding of vacuum energy and equilibrium dynamics.

He had moved beyond philosophy—this was now a scientific model waiting to be tested.

Kartik closed his notebook.

He was no longer just searching for connections—he was building a bridge between two ways of knowing.

And this time, he had the tools to make it undeniable.

The Unexpected Convergence

Kartik sat before his computer, running the latest set of simulations. The refined mathematical model—built with insights from control theory and structured energy interactions—was now producing data that was both measurable and repeatable.

And that's when he saw it.

A pattern.

Not just a random fluctuation, not just an anomaly, but a structured equilibrium forming within the system—self-correcting, self-sustaining.

He stared at the screen, then flipped through his notes on Rta.

"Rta is the rhythm of the cosmos, adjusting, balancing, ensuring harmony in all things."

It had sounded like philosophy before. But now, he saw it for what it truly was—a description of self-regulating dynamics in structured energy fields.

The realization sent a chill through him.

The ancient texts were not giving equations—they were describing observations.

Centuries ago, when these words were spoken, the sages did not have laboratories, interferometers, or mathematical proofs. But they had perception, introspection, and an awareness of nature's patterns.

And what they saw, they encoded—not as rigid formulas, but as experiential truths.

Science Without Instruments

Kartik leaned back in his chair, rubbing his forehead. He had spent years dismissing ancient wisdom as poetic mysticism. But what if it wasn't?

What if it was simply another way of recording knowledge—one that modern science had never learned to read properly?

Vedic thought did not prove truths through numbers—it experienced them, encoded them in metaphor, in rhythm, in verse.

And now, with the right framework, modern science was beginning to measure what had once been only observed.

Not different truths.

The same reality—seen through two different lenses.

Kartik took a deep breath, feeling the weight of what he had uncovered.

This was no longer just about aligning Vedic thought with physics.

It was about realizing that knowledge had always existed in multiple forms—and modern science was only now catching up.

A Scientist's Struggle

Kartik stared at the oscillating interference patterns on his screen, the rhythmic stabilizations that defied conventional expectations. The data was real. The experiment had been controlled, repeated, verified.

Yet the implications were anything but conventional.

He had spent his entire career anchored in empirical science, in measurable truths that could be tested and validated. He had never been one to entertain speculation, let alone draw connections between ancient texts and modern physics.

And yet...

The deeper he went, the harder it became to ignore the undeniable alignments between Vedic insights and structured energy dynamics.

He exhaled sharply, rubbing his temples.

If he continued down this path, it would challenge everything he had built his reputation on.

The scientific community thrived on skepticism—he had seen what happened to researchers who strayed too far from the accepted frameworks. He wasn't ready to be labeled as someone chasing mystical ideas.

Yet, at the same time, he couldn't dismiss the results.

He glanced at his notebook, filled with equations on self-regulating energy fields, stability feedback loops, and structured fluctuations—all concepts that aligned eerily well with descriptions of Rta in Vedic texts.

If he stopped now, he would be doing exactly what he had criticized others for—rejecting knowledge simply because it didn't fit the existing paradigm.

His choice was clear:

- Dismiss it, preserve his credibility, and move on.
- Continue testing, refining, and uncovering what no one else was willing to explore.

He clenched his jaw.

A scientist's duty was not to conform—it was to question.

And for the first time in his life, he knew exactly what he had to do.

He had to keep going.

A New Path Forward

Kartik sat in his office, staring at the neatly organized columns of data on his screen. His notebooks—once filled with skeptical questions—were now filled with structured observations, mathematical formulations, and carefully extracted insights from the Vedic texts.

But this was no longer about proving that ancient texts contained scientific knowledge.

That approach had been flawed from the beginning.

Instead, he now saw them for what they truly were—a guiding framework, a way of thinking about reality that modern science had overlooked.

Vedic philosophy had never been about proving anything in the way science demanded. It was about experiencing, understanding, and describing reality through direct observation and introspection.

Science, on the other hand, sought to quantify, to break down, to measure.

Neither approach was wrong. And perhaps, they were never meant to be separate at all.

Kartik took a deep breath. He had been trying to bridge these two worlds—to validate one using the other.

But that wasn't the answer. The answer was to use both, to allow them to complement each other, to fill the gaps where the other fell short.

- Where Vedic thought lacked precision, science could bring structure.
- Where science struggled with paradoxes and anomalies, Vedic insights could offer new ways of thinking.

This was no longer about ancient wisdom versus modern physics. It was about uncovering deeper truths—truths that had always existed, waiting for the right mind to connect them.

Kartik closed his notebook, feeling something he hadn't felt in a long time.

Not just curiosity. But purpose. This was just the beginning.

Part III — Consciousness As The Missing Variable

Reconciling the mind's presence in the equation of reality

Chapter 6: The Consciousness Conundrum

Kartik had spent weeks analyzing patterns in energy, structured equilibrium, and the principles of Rta. He had mapped the parallels between ancient philosophy and modern physics, uncovering alignments that challenged his understanding of reality.

But now, he had arrived at a far greater mystery—one that neither physics nor philosophy had fully explained.

What is consciousness?

It wasn't like the other forces he had studied. Gravity, electromagnetism, and even the underlying structure of space—all could be measured, defined, and tested.

But consciousness?

It was the only thing that could not be observed from the outside.

Science had dissected the brain, mapped neurons, and traced electrical impulses. Yet, no experiment had ever uncovered the source of awareness itself.

The Vedic texts, however, spoke of consciousness as the foundation of reality—not a byproduct, but the very fabric of existence.

This wasn't just a difference in interpretation—it was a difference in perspective.

Was the self an illusion of neural processes?

Or was it something deeper—a field, a force, a fundamental property of the universe itself?

For the first time, Kartik wasn't just questioning physics—he was questioning the nature of his own existence.

The Mind Beyond Matter

Kartik leaned back in his chair, flipping through the latest research papers on neuroscience and consciousness studies. His desk was cluttered with books—some modern, filled with brain scans and cognitive models; others ancient, speaking of the self, the observer, and the eternal nature of awareness.

The contrast between the two could not have been starker.

Modern science approached consciousness from the outside in—it dissected neurons, tracked synaptic firings, and analyzed how sensory inputs created perception. The prevailing view was simple: Consciousness was an emergent property of the brain.

But the Vedic texts saw it differently.

According to them, consciousness was not created—it was fundamental.

He glanced at a passage from the Brihadaranyaka Upanishad that he had underlined earlier:

"The self (Atman) is not born, nor does it die. It does not arise from the brain, nor is it bound by the body. It pervades all things, yet remains beyond perception."

If this was true, then consciousness was not a product of neural activity—it was something that existed independently, merely interacting with the brain.

Kartik frowned. That went against everything neuroscience claimed.

According to the materialist model, consciousness was an illusion—an elaborate trick created by electrical impulses and biochemical reactions. The brain processed reality, and awareness was simply an aftereffect.

But something about that explanation had always felt incomplete.

- If the brain created consciousness, then who was the "observer" experiencing it?
- If thought was just a collection of neurons firing, then why did self-awareness feel so distinct from mechanical processing?
- If perception was just an evolved function, then why did humans have the ability to reflect on existence itself?

He had read theories trying to explain this paradox—Integrated Information Theory (IIT), Global Workspace Theory, quantum models of the mind—but none of them truly solved the problem.

Because at its core, science was trying to measure something that might not exist in space and time the way physical objects did.

Kartik exhaled, staring at his notes.

If consciousness wasn't just an effect of the brain, then what was it?

A field? A force? A fundamental property of reality?

Or had modern science been asking the wrong question all along?

The Atman and the Universal Connection

Kartik turned the pages of the Upanishads, his fingers tracing the ancient Sanskrit verses. He had spent the last few weeks wrestling with the scientific view of consciousness—as a mere byproduct of neural activity—and the Vedic perspective, which claimed that awareness was not just an illusion of the brain but the very foundation of existence.

Now, he had arrived at a deeper distinction, one that unsettled him more than anything else so far.

The Vedic texts did not just speak of individual consciousness. They described something much greater:

Atman – The individual self, the pure awareness within each person.

Brahman – The universal consciousness, the underlying reality of everything.

He flipped to a passage from the Chandogya Upanishad:

"Tat Tvam Asi—That Thou Art."

The text collapsed the boundary between the self and the cosmos. It declared that the awareness within an individual was not separate from the universal consciousness—it was merely a reflection of it.

Kartik leaned back, rubbing his forehead. This was profound.

If true, it suggested that consciousness was not confined to individual minds, but rather an interconnected field—one that all living beings tapped into.

He frowned, staring at his notes.

This sounded eerily similar to something he had encountered before.

Modern physics had spent decades searching for a unified field theory—a fundamental force or medium that connected everything in existence. Einstein had pursued it, quantum physicists had theorized about it, and yet, it remained elusive.

Now, the Vedas were presenting an almost identical idea—but instead of a physical field, they described it as consciousness itself.

Could it be possible?

Had the ancient sages intuitively understood something that modern science was still struggling to define?

His scientific mind resisted the thought. Consciousness, by all current models, was an emergent property of complex neural networks.

But if that were true, why did these ancient texts describe it as something fundamental—something that existed beyond the physical?

And more importantly...

What if modern science had been looking for the wrong kind of field all along?

Neuroscience and the Illusion of Self

Kartik scrolled through the latest neuroscience papers on his screen, his eyes scanning diagrams of neural pathways, brainwave activity, and computational models of consciousness. Everything he read reinforced the dominant materialist view:

- Consciousness emerges from neural activity.
- The brain processes sensory input, generates thoughts, and creates a sense of self.
- There is no independent "I"—just an illusion formed by the brain's computations.

But something about this explanation felt incomplete.

He turned to one of the classic neuroscience experiments—Benjamin Libet's work on free will. The study had shown that the brain makes decisions milliseconds before a person becomes aware of them. The implication? Consciousness was just a passive observer, watching decisions happen rather than making them.

If that were true, then who was experiencing it all?

If the brain was just a machine processing data, then who was the "observer" behind it?

Kartik felt a familiar tension growing within him.

He had encountered this paradox before—but now, with the Vedic concept of Atman and Brahman in his mind, it took on an entirely new meaning.

In Vedic thought, consciousness was not produced—it was received.

"The self (Atman) is not created by the brain, nor bound by it. It is the witness of all things, yet remains untouched."

—Brihadaranyaka Upanishad

The brain, in this view, was not the source of awareness, but a tool for accessing it.

Like a radio receiving a broadcast, the brain translated consciousness into thought and perception—but the signal itself existed independently.

Kartik sat still, his pen hovering over his notes.

Could it be that modern science had been asking the wrong question?

Instead of asking:

"How does the brain create consciousness?"

Should they be asking:

"What is the brain interacting with to produce awareness?"

If consciousness was not a side effect of neurons, but something fundamental—a field, an essence, a force woven into the fabric of reality—then modern science had been looking in the wrong place all along.

And if that were true...

Then the Vedic texts had described it thousands of years before science had even realized the question existed.

The Quantum Puzzle of Awareness

Kartik sat across from the Acharya in the dimly lit temple courtyard, the night air cool against his skin. The rhythmic hum of distant chants faded into the background as the Acharya spoke.

"You have spent your life studying the universe as an outsider, Dr. Varun. Measuring it, breaking it down, searching for patterns." His voice was calm, deliberate. "But what if the universe is not something you observe? What if it is something you participate in?"

Kartik exhaled, leaning back. This was the very question that had unsettled him for weeks.

Modern physics had already stumbled upon a strange paradox—the act of observation itself could change the outcome of an experiment.

It was best demonstrated in the double-slit experiment—where particles behaved as waves when unobserved, but as discrete particles when measured.

This meant that consciousness—or at the very least, the act of observation—was somehow entangled with reality itself.

"Are you suggesting that consciousness is not just a process within the brain, but something that influences the external world?" Kartik asked.

The Acharya smiled. "That is not my suggestion. That is the nature of existence."

He turned the pages of an ancient manuscript and pointed to a passage:

"The observer and the observed are not separate. As one perceives, so the universe responds. The mind does not see the world; it shapes it."

Kartik felt a shiver crawl up his spine. This was not mere mysticism—it was eerily similar to what quantum mechanics was struggling to explain.

He had always dismissed the notion that perception could affect reality. But if the most fundamental particles in existence could shift their behavior based on whether they were being observed...

Then who—or what—was truly shaping reality?

The Acharya closed the manuscript and met Kartik's gaze.

"You believe that consciousness is an emergent phenomenon of the brain. But tell me, Dr. Varun—if a particle does not have a fixed state until it is observed, then is the observer not as fundamental as the particle itself?"

Kartik swallowed.

Science had spent decades trying to explain away the role of the observer in quantum mechanics, treating it as a mere mathematical formality.

But what if the ancients had already understood what science had yet to fully grasp?

Not that the brain generated consciousness—but that consciousness was the very fabric upon which reality was woven.

And if that were true, then the entire foundation of modern physics had been incomplete all along.

Beyond the Observer: The Self as the Field

Kartik sat alone in his study, the glow of his laptop screen casting sharp shadows across the room. His mind was restless, tangled in a thought he could neither fully accept nor entirely reject.

What if consciousness wasn't just an observer?

Physics had long accepted that the act of observation could alter reality—quantum mechanics had demonstrated that much. But mainstream science still treated this as a technicality, a problem to be resolved rather than a truth to be explored.

He flipped back to the Upanishads, where he had marked a passage:

"As the web is woven within the spider, As fire exists hidden within wood, As the sun is reflected in a thousand waters, So does the Self pervade all things."

Kartik tapped his fingers against his desk.

This was not describing the self as an observer, standing apart from reality. This was describing the self as the very fabric of existence.

The ancient texts did not just claim that consciousness observed reality—they suggested it was reality itself.

The Consciousness Field Hypothesis

In modern physics, everything was understood within the framework of fields.

- Gravitational fields dictated how objects moved in space.
- Electromagnetic fields determined forces between charged particles.
- Quantum fields underlay the existence of all matter.

But what if consciousness was also a field?

Not just a product of neurons, not just an emergent phenomenon—but the very substrate upon which reality was constructed.

Could it be that the Vedic concept of Brahman—the universal consciousness—was an early recognition of this fundamental field?

Kartik's mind raced. If science had been searching for a unified field theory to explain the universe, had it unknowingly been missing the most fundamental field of all—the field of awareness itself?

This was more than just philosophy now. It was an equation waiting to be written.

He leaned forward, his skepticism battling his intuition.

Had the ancient sages of the Vedas perceived what modern physics had only begun to suspect?

And if so, how had they understood it—without instruments, without mathematics, without experiments?

The answer, Kartik realized, would change everything.

Bridging the Intellect and Experience

Kartik sat across from the Acharya, the dim temple lamps flickering softly in the night air. His notebook lay open beside him, filled with equations, observations, and questions that refused to settle.

He had spent weeks dissecting the idea of consciousness as a fundamental field, mapping it against known physics, searching for a way to frame it in measurable terms. But no equation seemed enough.

The Acharya watched him in silence before speaking.

"You are searching for proof, Dr. Varun."

Kartik sighed. "That's how knowledge works. If something is real, it should be measurable."

The Acharya tilted his head slightly. "And how do you measure your own awareness?"

Kartik opened his mouth to respond but hesitated.

He could observe brain activity on a scan, measure neural responses to stimuli, even map the electrical signals of thought. But none of that explained the subjective experience of being conscious.

"You see," the Acharya continued, "science measures everything as an external object. But tell me—how can the mind measure itself?"

Kartik frowned. "There has to be a way."

"Perhaps," the Acharya said. "But would you ask a flame to illuminate itself?"

Kartik exhaled sharply, rubbing his temples.

This was the problem, wasn't it? Science had built a system of measurement that worked for everything—except for the one thing that made measurement possible in the first place.

If consciousness was the field that held reality, then the tools used to study physical forces would never be enough. It required a different approach.

The Vedic texts had claimed that consciousness was not an object, not a process, but the very essence of experience itself. And if that were true, then perhaps understanding it required something more than just intellect.

"Some truths," the Acharya said, "are not meant to be studied from the outside. They must be experienced."

Kartik stared at him. "You're saying I can't understand this unless I...what? Meditate?"

The Acharya smiled. "You have always sought truth through analysis, through reason. But tell me—have you ever sought truth through direct experience?"

Kartik wanted to dismiss the idea.

But a quiet voice inside him asked:

What if some knowledge cannot be proven—only realized?

And that thought unsettled him more than anything else.

Consciousness as the Key to Reality

Kartik had spent his life believing that reality existed independently of the observer. That the universe was governed by laws—fixed, measurable, indifferent to human perception.

But now, as he sat across from the Acharya, the foundations of that belief were beginning to crack.

"Everything you have studied," the Acharya said, "tells you that the universe is built from matter and energy. But have you considered that it is built from awareness?"

Kartik frowned. "You mean consciousness?"

The Acharya nodded. "Not as something confined to the brain, but as the fabric of reality itself."

Kartik inhaled sharply. This was an unsettling idea—but not an unfamiliar one.

In physics, some of the most cutting-edge theories suggested that consciousness might play a fundamental role in the structure of reality.

- **Quantum Mechanics** – The observer's presence collapses probability into reality.
- **Integrated Information Theory (IIT)** – Consciousness is not just emergent, but fundamental.

- **Theories of a Conscious Universe** – Some physicists had even proposed that the universe itself is aware at some level.

If consciousness wasn't created by the brain, but instead existed as a field, like gravity or electromagnetism, then everything changed.

Kartik rubbed his temples. "You're saying consciousness isn't just a byproduct of evolution—it's the underlying principle of existence?"

The Acharya smiled. "I am not saying it. The Upanishads have said it for thousands of years."

He flipped through the pages of an ancient text and pointed to a verse:

"Brahman is consciousness. The universe arises from it, is sustained by it, and dissolves into it."

Kartik exhaled.

He had always assumed physics and philosophy were separate. But now, he wasn't sure anymore.

Modern science was still debating the role of consciousness.

But the Vedic texts had already declared it the foundation of reality itself.

And for the first time, Kartik was beginning to see why.

The Dilemma of a Scientist

Kartik sat alone in his study, staring at the blinking cursor on his screen. His research notes lay scattered across the desk—neuroscience models, quantum mechanics equations, and verses from the Upanishads. The more he tried to organize his thoughts, the more they resisted structure.

His rational mind told him to walk away.

This was dangerous territory. Consciousness had always been treated as a philosophical question, not a scientific one. His entire career had been built on empirical inquiry, on hard data, on things that could be measured, tested, and reproduced.

And yet...

The deeper he went, the more he saw gaps in modern science—gaps that Vedic philosophy seemed to fill.

- Neuroscience could describe how the brain processed information, but it could not explain who was experiencing it.
- Quantum mechanics suggested that observation influenced reality, but it had no answer for why consciousness played a role at all.
- Physics was searching for a unified field theory, yet ancient texts had already described an all-pervading essence—Brahman, the fundamental awareness underlying existence.

Kartik clenched his fists.

Could it be possible?

Could consciousness be the missing equation—the fundamental field that science had yet to acknowledge?

If he accepted that, then everything changed.

The nature of reality, the definition of existence itself—it all had to be rethought.

But if he dismissed it, if he let his fear of academic scrutiny hold him back, then he was no better than the very people who refused to ask the difficult questions.

He took a deep breath.

Science had always progressed by challenging assumptions, by venturing into the unknown.

And if there was even the slightest chance that the ancients had glimpsed something modern physics was only now beginning to suspect—then he had to keep going.

Even if it meant questioning everything he once believed.

A Shift in Perspective

Kartik closed the last of his notebooks, leaning back in his chair. The weight of his own thoughts pressed against him. He had spent his entire career searching for answers within the rigid framework of science, believing that truth could only be found in what was measurable, in what could be proven.

But now, for the first time, he wasn't so sure.

He wasn't ready to accept everything the Vedic texts claimed. He wasn't convinced that ancient wisdom had solved the mysteries of the universe before science even began its journey.

But he could no longer dismiss the possibility that they had seen something real—something modern science had overlooked or forgotten.

The more he studied, the more it became clear that these texts were not primitive beliefs or philosophical poetry.

They were observations—structured, deliberate, and deeply insightful.

Had the ancients glimpsed truths that science was only now beginning to rediscover?

Had they described consciousness, not as a byproduct of the brain, but as the fundamental force underlying reality?

He didn't have the answers yet.

But for the first time, he wasn't looking for reasons to reject the questions.

There was something here—something worth exploring, worth testing, worth understanding.

And that was all that mattered.

He exhaled, feeling an unfamiliar sense of clarity.

Kartik had begun this journey as a skeptic.

But now, he was a seeker.

And he had only just begun.

Part IV — The Intellectual Confrontation

Facing the world that resists the unfamiliar

Chapter 7: Facing the Academic Gauntlet

Kartik stared at the polished surface of the conference table, his fingers tracing invisible patterns along its edge. His research notes lay neatly stacked beside his laptop, the weight of them both reassuring and suffocating.

This was it. The moment of reckoning.

For weeks, he had immersed himself in experiments, data analysis, and philosophical inquiries that had led him to an unsettling yet undeniable realization—consciousness, energy, and structured equilibrium were not separate forces, but interconnected aspects of reality.

But now, he had to take that realization out of the safe confines of his study and place it under the harsh scrutiny of the scientific community.

Would they see what he had seen? Or would they dismiss it outright?

He had spent his life believing that science was driven by curiosity, by the relentless pursuit of truth.

But today, he wasn't just presenting research.

He was challenging a paradigm.

And he knew all too well what happened to those who did.

Preparing for the Battle

Kartik sat at his desk, the glow of his screen illuminating the stacks of research papers, notebooks, and printouts scattered around him. This wasn't just another academic presentation—it was a test of how far he could push the boundaries of scientific inquiry without stepping outside its accepted framework.

He knew how fragile credibility could be. If he introduced the Vedic connections too early, he risked his entire argument being dismissed as mystical speculation. If he held them back too long, he might lose the deeper significance of his findings.

The balance had to be precise.

Structuring the Argument

He divided his presentation into three key sections, carefully controlling how the ideas would unfold:

Experimental Data First: He would begin with hard evidence—measurable, reproducible results. He had structured energy fluctuations, patterns that pointed to an intrinsic self-regulating field. No mention of philosophy yet—just physics.

Theoretical Implications: He would then compare his findings with existing models in vacuum energy, self-organizing systems, and quantum field dynamics. He would acknowledge gaps in modern physics, showing how his results raised new questions.

Bridging Ancient Insights: Only after establishing scientific credibility would he introduce Vedic correlations—not as validation, but as an alternative perspective that had anticipated similar ideas. He would position Rta and structured consciousness as guiding principles, not mystical concepts. His goal was not to prove the Vedas scientifically, but to show that ancient thought and modern science might be describing the same truths in different ways.

Anticipating the Criticism

He knew the objections that would come. His critics wouldn't attack his data—they would attack the implications.

"Correlation is not causation." – He prepared counterarguments, emphasizing that patterns in independent systems often indicated deeper principles yet to be discovered.

"This sounds like metaphysics, not science." – He would ground every claim in observable phenomena rather than philosophical speculation.

"What practical use does this have?" – A question he had anticipated. If consciousness was a fundamental force, understanding it could revolutionize everything from artificial intelligence to medical sciences.

As he refined his slides, he realized something.

He wasn't just defending his research—he was defending the very idea that knowledge did not belong to one era, one civilization, or one method of inquiry.

The truth, wherever it came from, deserved to be explored.

He saved the final draft of his presentation, took a deep breath, and closed his laptop.

He was ready.

Now, he just had to face the room.

The Challenge of Framing the Narrative

Kartik stared at his laptop screen, the glowing presentation slides reflecting in his tired eyes. This was the hardest part—not the research, not the data, but the way he would frame it.

He knew how fragile credibility was in academia. The moment a theory strayed from the accepted language of science, it risked being labeled pseudoscience, regardless of the data supporting it.

And now, he was walking that very fine line.

To Mention Vedic Thought or Not?

He had two options:

- **Let the data speak for itself**—avoid any mention of Vedic philosophy and focus purely on the structured energy dynamics he had observed.
- **Acknowledge the ancient parallels**—present the Vedic insights carefully, not as proof but as an intuitive framework that seemed to align with the results.

If he chose the first, his findings might be taken more seriously, but the deeper significance of what he had uncovered would remain unspoken.

If he chose the second, he risked immediate dismissal—even before the discussion began.

Finding the Middle Ground

After hours of internal debate, Kartik realized the answer wasn't either-or—it was how he introduced the ideas.

Data First, Interpretation Later: He would start with observable facts, reproducible results, and structured energy patterns—all grounded in experimental science. No mention of philosophy. No references to ancient texts.

Addressing the Gaps in Physics: He would highlight the paradoxes and anomalies that modern science struggled to explain. Instead of forcing conclusions, he would ask, "Is there a missing link?"

Introducing Ancient Perspectives as a Thought Experiment: Only after establishing scientific credibility would he introduce Vedic thought—not as proof, but as an alternative lens. He would frame it as a historical observation, not a validation of ancient texts. Instead of saying, "The Vedas knew this first," he would ask, "Could this be an intuitive insight into what we are only now beginning to quantify?"

This way, he wouldn't be pushing a claim—he would be inviting a discussion.

The Final Decision

Kartik sat back and sighed. This wasn't just about presenting research. It was about breaking the barrier between science and ancient wisdom—without losing credibility in either.

He glanced at the time. Tomorrow, he would stand before the scientific community.

His findings were solid.

His argument was refined.

Now, he had to see if the world was ready to listen.

A Divided Camp: Conversations with Colleagues

Kartik had expected skepticism. What he hadn't anticipated was how deeply divided his colleagues would be.

As word of his upcoming presentation spread, so did the quiet conversations—some in support, others in caution, and a few in outright dismissal.

Encouragement from the Open-Minded

In the physics department lounge, Dr. Ananya Roy, a theoretical physicist specializing in vacuum fluctuations, leaned forward, her eyes sharp with curiosity.

"This is bold, Kartik," she said, flipping through his notes. *"You're proposing that these structured energy fluctuations aren't random noise, but part of a deeper equilibrium?"*

Kartik nodded. *"That's what the data suggests. There's a stabilizing factor that isn't currently accounted for in mainstream physics."*

She smirked. *"And let me guess—the moment you mention ancient philosophy, people stop listening."*

He exhaled. *"Exactly. But the parallels are too strong to ignore. I'm not saying the Vedic texts predicted this. I'm saying they might have intuited something fundamental—something we are only now beginning to quantify."*

Dr. Roy tapped her pen against the table. *"Well, I think you should push forward. If the data is solid, the implications will stand on their own. And honestly?" She leaned back. "Theoretical physics could use a shake-up."*

Warnings from the Traditionalists

Not everyone shared her enthusiasm.

Later that evening, Kartik sat across from Dr. Verma, a senior physicist known for his rigid adherence to conventional models. The older man sighed, setting down his coffee.

"Kartik," he said, his voice calm but firm, *"you're walking into dangerous territory."*

Kartik folded his arms. *"Because I'm presenting new findings?"*

Dr. Verma shook his head. *"Because you're tying them to ancient knowledge. You know how this works. The moment you step outside the accepted framework, you're no longer a physicist—you're a 'fringe thinker.' And once that label sticks, it's hard to shake."*

Kartik clenched his jaw. *"So I should ignore valid insights just because they come from an unconventional source?"*

"I'm saying," Dr. Verma continued, *"that science only progresses when it stays within measurable, empirical methods. If you bring in philosophy, even indirectly, you risk losing credibility before the conversation even begins."*

Kartik exhaled. He understood the warning. But was staying silent the right approach?

A Moment of Reflection

As he walked back to his office, Kartik realized something.

He had support from those who valued exploration.

And he had warnings from those who feared stepping outside the academic status quo.

Neither side was entirely wrong.

But if he let fear dictate his path, was he really pursuing truth—or just protecting his reputation?

No. He had to present his findings.

If there was truth to be uncovered, then silence was not an option.

Constructing the Defense

Kartik sat in his office, staring at the draft of his presentation. He wasn't just preparing a talk—he was preparing for battle.

The skepticism he had encountered was predictable. He was challenging deeply entrenched scientific assumptions. That meant he had to ensure that his argument was airtight, not just persuasive but unassailable on scientific grounds.

Anticipating the Criticism

He listed the most likely objections and the counterarguments he would need:

Objection 1: "Your results are just an anomaly."

Counter: "The experiment has been repeated multiple times with consistent results. Furthermore, similar stabilization patterns have been observed in other complex systems—suggesting an underlying principle rather than a random occurrence."

Objection 2: "There's no theoretical framework to support your findings."

Counter: "Every major breakthrough in physics—from relativity to quantum mechanics—was first an observation before it was a theory. The absence of a framework does not invalidate experimental data. Instead, it suggests the need for a new model."

Objection 3: "Ancient texts have no place in physics."

Counter: "I am not claiming these texts are scientific theories. I am presenting them as historical perspectives that align with certain modern observations. Ignoring them simply because they are old is not scientific—it is a bias against knowledge from unconventional sources."

Objection 4: "This sounds too much like metaphysics."

Counter: "Many concepts that were once considered metaphysical—like the idea of a curved spacetime—are now fundamental to physics. We should not dismiss an idea just because it does not yet fit into our existing models."

Sharpening the Language

He knew he had to be precise in his wording. Any misstep could give critics an easy way to discredit his work.

Wrong Approach: "The Vedic texts describe the same principles as modern physics."

Refined Approach: "Certain descriptions in the Vedic texts align with observable scientific phenomena, offering a historical perspective on structured energy interactions."

The difference? One makes a claim. The other presents an observation.

Building Credibility

To strengthen his argument, he:

- Cross-referenced his findings with independent studies on self-organizing energy systems.
- Reached out to physicists who studied emergent

structures in vacuum energy.
- Compiled case studies where scientific breakthroughs had come from unexpected sources.

This wasn't about proving ancient wisdom was right. It was about asking whether modern science had overlooked something valuable.

Final Preparations

As Kartik rehearsed his presentation, he felt the weight of what was at stake.

This was more than just defending a theory. It was about expanding the boundaries of what science was willing to explore.

Tomorrow, he would step into that lecture hall not as someone seeking validation, but as someone demanding the right to question.

And he was ready.

The Pressure of the Scientific Establishment

Kartik sat in the dim glow of his office, the quiet hum of his computer the only sound breaking the stillness. His final presentation draft was ready, his defense well-prepared. Yet, a weight settled in his chest—a tension he couldn't ignore.

He had spent years believing in science as a pursuit of truth, an ever-evolving discipline where new discoveries refined or replaced old ones. But now, he had seen another side of it—the rigidity, the resistance, the unspoken rules about what was acceptable to question.

History had shown this pattern before.

Heliocentrism: When Copernicus proposed that the Earth revolved around the Sun, he was ridiculed. It took decades before the idea was accepted.

Germ Theory: Before Pasteur's experiments, the idea that invisible microbes caused disease was dismissed as nonsense.

Quantum Mechanics: When Planck and Einstein introduced ideas that defied classical physics, even their own peers resisted.

Kartik tapped his fingers against the desk. Was he standing at a similar crossroads?

His findings did not claim to rewrite physics—but they challenged existing models in ways that made people uncomfortable. And that was enough to invite skepticism, if not outright rejection.

He imagined the reactions he might face:

- Curiosity from some.
- Skepticism from most.
- Dismissal from those unwilling to question their foundations.

Was he ready for this?

He had dedicated his life to scientific integrity, and yet, here he was, on the verge of being labeled a fringe thinker—just for asking questions that science itself had yet to answer.

A part of him wondered if he should soften his approach, frame his findings in a way that conformed to mainstream ideas.

But then, another thought pushed back.

If science was truly about discovery, should it not welcome the difficult questions?

He had seen the data. He had run the experiments, refined the models, anticipated every counterargument.

The fear of professional scrutiny was real. But the fear of self-betrayal—of ignoring the truth because it was inconvenient—was greater.

He straightened his posture.

Tomorrow, he would step into that lecture hall.

Not just to defend his findings.

But to defend the right to question, to explore, and to push science beyond its comfort zone.

Doubt vs. Conviction

Kartik stood by his office window, watching the city lights flicker in the distance. His presentation was tomorrow, and for the first time, he wasn't sure if he was ready.

It wasn't the science he doubted. His data was solid, his methodology rigorous, and his argument well-structured. But would it be enough?

Would it be enough to convince an audience that had already decided what was acceptable to question?

The temptation to alter his approach crept in—perhaps if he toned down the discussion on consciousness, if he framed his findings purely within mainstream physics, he could avoid immediate dismissal.

He had spent years working within the system, understanding its rules. He knew how academia functioned—how it demanded conformity before it considered innovation.

But was that what he wanted?

Would sacrificing the full depth of his findings make them more acceptable, or would it make them meaningless?

He thought back to every breakthrough in science that had faced resistance, every time the establishment had fought against ideas that threatened its foundations.

- Galileo had been silenced for challenging geocentrism.
- Einstein's ideas on relativity had been dismissed before they reshaped physics.
- Even quantum mechanics had been labeled as absurd when first proposed.

And now, here he was—standing at the edge of his own paradigm shift.

This wasn't about proving the Vedas right or redefining science overnight—it was about asking the questions no one else was willing to.

If he backed down now, what was the point of everything he had uncovered?

Kartik took a deep breath.

He would not alter his findings to fit expectations.

He would not dilute the questions that needed to be asked.

Tomorrow, he would stand in that lecture hall and present the truth as he had found it—without hesitation, without fear.

Because if science truly sought knowledge, then it could not turn away when faced with something it did not yet understand.

The Moment of Truth

Kartik stepped into the conference hall, his heartbeat steady but his mind razor-sharp. This was it.

Rows of scientists, professors, and graduate students filled the auditorium, some leaning forward with curiosity, others sitting back with folded arms, their skepticism already apparent.

This was not just a presentation. It was a trial.

He made his way to the podium, plugging in his laptop, the projector screen flashing to life behind him. The title of his research appeared:

"Structured Energy Equilibrium and the Role of Consciousness in Physical Systems"

He had deliberately left out any direct mention of Vedic philosophy. The data had to speak first.

He cleared his throat and began.

The Evidence First

"For the past several months, I have been studying a specific pattern in structured energy fluctuations—one that does not align with conventional models of stochastic behavior."

He clicked to the next slide, displaying the experimental setup and the observed data.

"There is a stabilizing factor at play," he continued. "These fluctuations are not purely random. They exhibit self-regulation—something that current physics does not fully account for."

A murmur spread through the audience.

He flipped to another slide, showing comparative analysis.

"Furthermore, these findings parallel self-organizing behaviors seen in complex systems. Yet, the question remains—what governs this intrinsic balance?"

He paused, scanning the room. Now came the hard part.

Introducing the Unconventional Perspective

"If we consider historical perspectives on the nature of balance and order, we find intriguing correlations."

Click.

A verse from the **Rig Veda** appeared on the screen:

"Rta—the unseen order that sustains all things, Governing what moves and what does not, Holding the rhythm of the universe."

There it was. The shift in the room was immediate.

Some attendees leaned forward, intrigued. Others stiffened, their skepticism palpable.

Kartik held his ground. *"I am not suggesting that ancient texts were scientific models. But I am suggesting that perhaps, embedded in these observations, was an intuitive understanding of equilibrium that aligns with what we are only now beginning to quantify."*

The Expected Challenge

A hand shot up. Dr. Mehta. Of course.

"This is a fascinating study," the older physicist said, his tone carefully measured. "But are you implying that ancient philosophy anticipated structured energy dynamics?"

Kartik met his gaze. "Not in a predictive sense. I am suggesting that human understanding has always evolved through different modes of inquiry. Science quantifies. Ancient philosophy observed. If both lead to similar conclusions, then perhaps we should be open to re-examining ideas from multiple perspectives."

More murmurs. Some nods. Some smirks.

Dr. Verma, another senior physicist, leaned back. "Correlation is not causation, Dr. Varun."

"I agree," Kartik said smoothly. *"Which is why I am not asking you to accept this perspective—I am asking if we are willing to explore what we have overlooked."*

Silence.

For the first time, no immediate rebuttal.

The Turning Point

He flipped to his final slide.

"If science is truly a pursuit of knowledge, then should we ignore possible insights simply because they come from unconventional sources?"

Kartik let the question hang in the air.

He had not come to convince them.

He had come to make them think.

And as he stepped away from the podium, he knew—that was exactly what he had done.

Chapter 8: The Symposium Showdown

The auditorium was larger than Kartik had expected. Rows of polished seats stretched out before him, filled with some of the sharpest minds in physics, neuroscience, and philosophy. The air buzzed with quiet conversations, the kind that carried an undertone of skepticism.

He had presented his research before, but this was different.

This wasn't just an academic conference—this was a global symposium, a stage where reputations were made or destroyed.

Some had come out of genuine curiosity, intrigued by the bold nature of his findings. Others were here for a different reason—to dissect, to challenge, to remind him where the boundaries of science truly lay.

Kartik adjusted his collar, steadying his breath. He had prepared for this.

The data was solid. The arguments were refined.

But today wasn't just about defending his research.

Today was about challenging the very framework of scientific inquiry itself.

An Invitation to the Global Stage

The email had arrived late at night, catching Kartik off guard. At first, he thought it was just another routine conference announcement, one of the many that flooded his inbox. But as he read further, his pulse quickened. It was an official invitation to present his research at the International Symposium on Physics and Consciousness, a high-profile event where some of the world's leading minds gathered to debate the intersection of quantum mechanics, neuroscience, and the nature of reality.

He read the email twice, then a third time, almost expecting to find a mistake. This was not just another academic gathering—it was a platform where controversial ideas were debated, dissected, and often dismissed outright. The symposium had a reputation for drawing two types of participants: those genuinely interested in expanding the boundaries of human knowledge, and those who came solely to tear apart anything that deviated from conventional scientific models.

This was an opportunity. A chance to take his work beyond the confines of his academic circle, beyond private discussions and departmental meetings. But it was also a risk. Until now, he had only shared his findings with select colleagues, trusted minds who either encouraged his work or, at the very least, engaged with it critically but constructively.

The symposium, however, would be different. Here, he would be speaking in front of the broader scientific community, including those who had spent their careers defending the boundaries of established science. Some of them would listen with curiosity, intrigued by the bold nature of his research. Others would be there for a different reason—to challenge, to discredit, and to remind him where the limits of scientific inquiry were supposed to lie.

He leaned back in his chair, rubbing his temples. This wasn't just about presenting his work—it was about facing the scientific establishment head-on. He had no illusions about what that meant. Science, for all its claims of objectivity, was not immune to resistance, to dogma, to the inertia that slowed the acceptance of radical new ideas. History had proven that again and again.

He had two options. He could decline the invitation, keep refining his work in private, and wait for the right moment—whenever that was. Or he could step onto that stage, knowing full well that the moment he spoke of consciousness in the context of structured energy equilibrium, skepticism would harden, and criticism would sharpen.

But wasn't this why he had pursued this path in the first place? To challenge assumptions, to ask the difficult questions, to push science beyond its comfort zone?

He exhaled slowly and clicked the reply button.

He accepted the invitation.

Whatever happened next, he was ready to face it.

A High-Stakes Presentation

Kartik stepped onto the stage, his shoes clicking softly against the polished floor. The auditorium was packed—a rare blend of physicists, neuroscientists, and philosophers, all gathered to challenge the boundaries of scientific thought. Some watched him with polite curiosity, others with skepticism barely concealed behind neutral expressions.

He adjusted the microphone and took a steady breath. This was the moment.

For weeks, he had refined his presentation, knowing that one misstep—one phrase too suggestive, one analogy too bold—could turn the discussion against him. He couldn't afford to let this become a debate about mysticism versus science. His findings had to stand on their own.

With a click, the first slide appeared behind him.

"Structured Energy Equilibrium and the Role of Consciousness in Physical Systems."

No mention of Vedic philosophy. No reference to ancient wisdom—not yet. He had deliberately crafted his introduction to speak the language of physics, letting the data lead the way.

"As we deepen our understanding of energy fields and self-organizing systems," he began, "there remain certain anomalies in how equilibrium is maintained within structured environments."

He flipped to the next slide—graphs, waveforms, recorded fluctuations from his experiments.

"This data," he continued, "demonstrates a stabilizing factor at play—one that prevents complete entropy collapse, yet does not fit within the conventional models of stochastic behavior. What we are observing here is not just a pattern, but a self-regulating principle inherent to the system itself."

A few murmurs rippled through the audience. Good. They were engaged.

He continued, carefully building his argument:

- The patterns in energy fluctuations were not random but exhibited an intrinsic equilibrium.
- Existing theories in quantum field dynamics and vacuum fluctuations did not fully explain the observed self-regulation.
- If a stabilizing force existed at a fundamental level, what was its origin? What governed it?

He paused, letting the question settle.

Now, the real challenge began.

With another click, a new slide appeared—a passage from an ancient manuscript.

Introducing the Historical Perspective

Kartik knew this was the moment that would decide everything.

He turned to face the audience, carefully modulating his tone.

"What's interesting," he said, "is that this idea—that equilibrium is not simply a function of randomness, but a principle that sustains order—has been contemplated before. Not in scientific terms, but in a different mode of inquiry."

He gestured toward the screen. A passage from the Rig Veda appeared:

"Rta—the unseen order that sustains all things, Governing what moves and what does not, Holding the rhythm of the universe."

A shift in the room. The skeptical expressions deepened. He had anticipated this.

"I want to be clear," he continued before anyone could interject, *"that I am not presenting ancient texts as scientific models. What I am suggesting is that across different eras and different methods of thought, humanity has attempted to describe the same fundamental principles."*

Some heads nodded slightly. Others remained rigid, waiting to challenge him.

He moved to the next slide, reinforcing his argument:

- Modern science quantifies reality.
- Ancient wisdom conceptualized it through direct

experience.
- Both describe the same patterns—one in equations, the other in philosophy.

"This," Kartik concluded, *"is not an attempt to validate ancient texts through science or to turn science into philosophy. This is an attempt to ask a simple question: Are we willing to expand our methods of inquiry if they lead us to deeper truths?"*

Silence. Not rejection, not yet—but not immediate acceptance either.

He had laid the foundation.

Now, the real test would begin in the Q&A.

The Core Argument: Science and Spirituality as Complementary Paths

Kartik took a measured breath, glancing at the sea of faces before him. The auditorium was still, but he could feel the weight of skepticism pressing against him. This was the moment where his argument would either hold ground or be dismissed outright.

He clicked to the next slide, displaying a comparative analysis. On one side—his experimental findings on structured energy equilibrium. On the other—a historical excerpt from the Vedic texts describing Rta, the cosmic principle of balance.

"I want to emphasize," he began, *"that my research does not seek to replace scientific inquiry with philosophical speculation. Instead, I am asking whether we are limiting our scope by ignoring perspectives that hint at deeper truths."*

He turned to the screen.

"We have observed structured equilibrium in energy fluctuations, a self-regulating mechanism that maintains stability. Existing

physical models struggle to explain the origin of this balance. What we see here," he gestured to the Vedic passage, *"is a description—not in mathematical terms, but in conceptual understanding—that speaks of an unseen order governing what moves and what does not."*

He paused, scanning the audience. Some nodded slightly, intrigued. Others remained rigid, waiting to challenge him.

"We must recognize," Kartik continued, *"that science and spirituality have always approached knowledge differently. Science is a process of quantifying reality, reducing it to equations, measurements, and testable predictions. Ancient philosophy, on the other hand, sought to experience reality directly, to understand it through introspection and observation. These are not competing methods—they are complementary."*

He shifted the slide again.

"When we look at the evolution of scientific thought, we see that many breakthroughs have emerged from ideas once considered purely philosophical. The notion of atoms, the curvature of space-time, even the probabilistic nature of reality in quantum mechanics—all of these ideas once existed in speculative thought before science developed the means to quantify them."

He stepped away from the podium slightly, his voice firm but steady.

"So, I ask—are we, in our pursuit of knowledge, limiting science by disregarding insights that were arrived at through different means? If certain philosophical traditions have described principles that align with modern physics, does it not warrant deeper investigation?"

Silence. Some faces reflected intrigue. Others were already preparing counterarguments.

Kartik knew he had walked the fine line carefully—introducing Vedic thought without overstepping into unverifiable claims, positioning it not as an alternative but as a parallel inquiry.

But would that be enough?

He took a breath.

Now, the real battle would begin.

The First Wave of Resistance

The silence in the auditorium barely lasted a few seconds before the first hand shot up. Kartik recognized Dr. Alan Whitmore, a well-respected theoretical physicist known for his sharp critiques of unconventional theories. His reputation alone was enough to command attention in any academic gathering.

"This is an interesting presentation, Dr. Varun," Whitmore began, his voice measured but carrying an unmistakable undertone of challenge. *"But what you're describing—these structured energy fluctuations—how can you be certain that they represent something fundamental and not just a localized anomaly within your experimental setup?"*

Kartik had expected this. He knew that before anyone even considered his deeper implications, they would attack the validity of his data itself.

He nodded, maintaining a calm demeanor. *"That's a fair question. But the data is not a one-time occurrence. The patterns have remained consistent across multiple trials under varying conditions. I've controlled for external interferences, tested different configurations, and the results persist."*

Whitmore leaned back slightly, unimpressed. *"But that still doesn't explain their origin. Just because a pattern emerges doesn't mean it reveals a deeper law of physics. It could be an unknown environmental factor, a limitation of instrumentation, or even an observer bias in the way you're interpreting the data."*

Kartik anticipated this too. *"Absolutely. That's why I'm not making any definitive claims yet. But I do believe these findings suggest that there's a self-regulating mechanism at play—one that deserves further exploration rather than immediate dismissal."*

He let the statement hang in the air. Whitmore didn't respond immediately, but his expression remained skeptical.

Before Kartik could continue, another voice cut in—this time, Dr. Emilia Torres, a physicist who had been vocal in her rejection of anything she perceived as pseudo-science.

"You're being careful with your wording," she said, her tone sharper than Whitmore's. *"But let's not ignore the subtext here. You're drawing connections between your research and ancient philosophy. Are we supposed to take Vedic texts as scientific sources now?"*

A murmur rippled through the audience. This was the real battle.

Kartik inhaled. This was the reaction he had feared—but also prepared for.

"I want to be clear," he said, firm but composed. *"I am not claiming that ancient texts are scientific theories. That would be an oversimplification."*

He paused, letting his words settle before continuing.

"But scientific inquiry has always been about recognizing patterns, even in places we might not expect. If a historical source describes a fundamental principle that aligns with observed physical phenomena, why should we ignore it simply because of its origin?"

Some nodded. Others folded their arms, unconvinced.

Dr. Torres wasn't done. *"Because there's a difference between poetic metaphors and empirical models. The texts you reference were written before modern physics existed. Any similarity is coincidental at best and confirmation bias at worst."*

Kartik held his ground. *"Then let's take that same logic and apply it across history. Ancient cultures speculated about atoms before microscopes could confirm their existence. Time dilation was a philosophical idea before relativity provided mathematical proof. The concept of an interconnected universe existed long before quantum mechanics suggested entanglement."*

He turned slightly, scanning the room. *"Do we dismiss those early ideas as well, simply because they weren't formalized in the language of modern science?"*

A beat of silence followed. It wasn't a victory, but it wasn't outright dismissal either.

The resistance had come, as he had expected.

But not everyone in the room was unconvinced.

And that was enough to keep going.

Kartik's Defense: Standing His Ground

Kartik took a slow breath, steadying himself. This was the moment that would decide whether the discussion moved forward or collapsed into outright dismissal.

He could feel the weight of skepticism in the room, but he had prepared for this. He wasn't here to defend ancient philosophy—he was here to defend the pursuit of knowledge, no matter its origin.

"I fully acknowledge the skepticism," he began, keeping his voice measured. *"In fact, I would be skeptical too if someone claimed ancient texts held scientific truths before we had the tools to measure them."*

Some members of the audience nodded slightly. Good. Acknowledging their doubt kept them engaged.

"But let me be clear," he continued. *"I am not here to prove that Vedic philosophy anticipated modern physics. I am not here to claim that ancient thinkers 'knew' about structured energy fields or quantum mechanics in the way we define them today. That would be an oversimplification."*

He paused, then gestured to his slides.

"What I am saying," he continued, more firmly now, *"is that certain historical perspectives seem to align with observed scientific phenomena—not because they are scientific theories, but because they may represent early intuitive understandings of reality."*

He moved to the next slide, showing a comparison between scientific models of structured equilibrium and Vedic descriptions of Rta.

"Consider this," Kartik said. *"Throughout history, knowledge has been preserved in different ways. Some through equations, some through metaphors, some through direct experience. What we call science today was once philosophy. And what was once dismissed as mysticism has, time and again, resurfaced in new ways through scientific discovery."*

He let the words settle. No one interrupted.

"So my question is this," Kartik said, scanning the room. *"Are we willing to explore the possibility that some knowledge—however it was framed—might have been preserved in ways we have not yet fully understood?"*

Dr. Whitmore, the theoretical physicist who had challenged him earlier, leaned back in his chair, thoughtful but unconvinced. *"Even if we grant that certain ideas resemble modern physics, resemblance is not the same as evidence. How do you differentiate between genuine insight and retrofitting ancient ideas onto contemporary science?"*

Kartik had expected this.

"That's the right question," he admitted. *"And I would argue that the answer is in the data. If the only thing we had were philosophical texts making vague claims, I wouldn't be standing here. But we also have experimental results that point to an underlying order in structured energy systems—an equilibrium we do not yet fully understand."*

He gestured to his experimental findings.

"My research stands on scientific merit alone. The historical parallels do not define it, they simply raise a deeper question—why do these similarities exist? Are they coincidence, or do they suggest a deeper layer of knowledge that modern science has yet to formalize?"

Another pause. This time, the resistance was not as firm.

He had shifted the conversation from proving ancient knowledge to expanding modern inquiry.

And for the first time, he saw hesitation, not rejection, in the faces before him.

The debate was far from over.

But at least the door had not been shut.

The Unexpected Support

Just as Kartik was preparing for another wave of skepticism, a voice from the middle of the auditorium cut through the tension.

"I have a question," said Dr. Aisha Patel, a cognitive scientist known for her work in consciousness studies and neural correlates of awareness. She leaned forward, curiosity evident in her expression. *"You mentioned that your findings suggest an underlying*

structure regulating energy fluctuations. Have you considered that something similar might apply to consciousness itself?"

Kartik blinked. This was unexpected.

She continued, *"In cognitive science, we still don't have a complete model for how subjective experience arises from neural activity. Most mainstream theories assume that consciousness is emergent—just a product of complex brain functions. But some recent studies suggest a field-like nature of consciousness, something that extends beyond the individual brain."*

A murmur spread through the audience. This was not the kind of response most were expecting.

Dr. Patel gestured toward the screen where Kartik had presented his findings. *"The idea that structured equilibrium exists in energy fields is interesting. But what's even more compelling is that the Vedic traditions describe consciousness in a similar way—not as something generated by the brain, but as something fundamental, which the brain interacts with."*

Kartik nodded, carefully choosing his words. *"That's exactly the kind of interdisciplinary approach I'm advocating for. Science has done an incredible job in mapping the brain's functions, but it has yet to explain the first-person experience of being aware. If we consider the possibility that consciousness is not just a byproduct of the brain, but an underlying field that the brain tunes into, then we may be looking at a missing layer in our understanding of reality."*

Dr. Patel smiled slightly. *"Which, of course, is a difficult position to argue in physics."*

A few chuckles broke the tension, but the shift in the room was noticeable.

Dr. Robert Chen, a physicist specializing in emergent systems, added, *"To be fair, interdisciplinary insights have led to some of the biggest breakthroughs in science. When neuroscience and artificial intelligence intersected, we gained entirely new perspectives on cognition. Perhaps this is another such moment."*

Not everyone was convinced, but the resistance was no longer absolute.

Even Dr. Whitmore, who had been Kartik's most vocal critic, seemed more engaged now than dismissive. *"It's an interesting angle,"* he admitted. *"It doesn't mean we accept the premise outright, but maybe there's something to investigate further."*

Kartik exhaled, allowing himself a small moment of relief. This was progress.

For the first time since he had stepped onto that stage, he wasn't the only one making the argument.

The idea had found a foothold.

And that meant the conversation wasn't over—it was just beginning.

The Q&A: A Clash of Perspectives

The moderator stepped forward, signaling the beginning of the Q&A session. Kartik gripped the edges of the podium, steadying himself. This was the real test.

A hand shot up almost immediately. Dr. Whitmore, again.

"You've demonstrated an interesting anomaly," he began, *"but anomalies in physics are only meaningful if they lead to a predictive framework. Do you have a working model that can mathematically describe this structured equilibrium?"*

Kartik had expected this. Without equations, physics remained unconvinced.

"I'm in the early stages of developing a model," he admitted. *"What I have observed is a recurring stabilization pattern that*

suggests an underlying principle of self-regulation in energy fluctuations. It is not yet a complete theoretical framework, but neither was quantum mechanics in its early stages."

A few nods. A few skeptical glances.

Another hand went up. Dr. Emilia Torres, the outspoken critic from earlier.

She adjusted her glasses. *"Even if we assume that your findings are valid, you are implying that consciousness plays a role in fundamental physics. That's a bold claim. Why should we entertain it when consciousness, to our knowledge, is just an emergent property of neural activity?"*

Kartik inhaled. This was a critical moment. He had to be precise—one wrong word, and the argument would collapse into speculation.

"I appreciate the skepticism," he said, maintaining a composed tone. *"To clarify—I am not claiming that consciousness directly influences physical reality in the way some interpretations of quantum mechanics suggest."*

He paused, then continued, *"However, what I am proposing is that consciousness may be more fundamental than we currently assume. If we are seeing stabilizing effects in structured energy fields, and if cognitive science is also questioning whether consciousness is merely emergent, then perhaps there is an underlying principle that connects them both. We don't need to redefine physics—only expand our perspective on what reality might be."*

A murmur spread across the room. That was the right balance.

Dr. Robert Chen, the emergent systems physicist who had shown earlier interest, leaned forward. *"So, in essence, you're not claiming that the ancients had a scientific model—only that their philosophical insights might have pointed to concepts we are only now beginning to measure?"*

"Exactly," Kartik replied. *"The Vedic texts did not provide equations, but they described fundamental patterns—balance, interconnectedness, structured reality—that seem to align with what we are now quantifying in physics. This is not about proving ancient texts right; it's about recognizing that different intellectual traditions may have observed the same truths in different ways."*

The room was no longer just challenging him. They were thinking.

Kartik stole a glance at the moderator. Time was almost up.

One last question came from an older neuroscientist in the back of the room.

"If what you're suggesting is true, then where does this leave science?"

Kartik exhaled, knowing this was the most important question of all.

"Science is a process of discovery," he said. *"Every field, from physics to biology, evolves by questioning its own assumptions. If structured equilibrium in energy is real, and if consciousness is more fundamental than we believed, then science doesn't have to abandon its principles. It simply has to expand them."*

The room fell silent for a moment.

The moderator stepped forward. *"Thank you, Dr. Varun. That concludes this session."*

Kartik nodded, stepping away from the podium.

He didn't know if he had changed minds today.

But he had done something just as important—he had forced them to reconsider.

A Divided Audience: The Aftermath

As Kartik stepped away from the podium, the air in the auditorium still buzzed with conversation. Some attendees were engaged in quiet but intense discussions, while others were already gathering their belongings, their expressions ranging from thoughtful intrigue to firm skepticism.

He had known this would happen. This was not the kind of presentation that ended with universal agreement—it was the kind that disrupted certainty.

Some scientists approached him with curiosity, eager to discuss his findings further. Dr. Robert Chen, who had shown a willingness to engage with his ideas, shook his hand.

"You've opened up an interesting discussion," Chen said. *"You might not convince everyone right away, but that's how every shift in thinking starts."*

Kartik nodded. *"I wasn't expecting to convince everyone. I just wanted to get them to consider the question."*

Chen smiled. *"Well, you've certainly done that."*

But not all reactions were encouraging.

Dr. Emilia Torres walked past him, muttering something to a colleague. Kartik caught only part of it—*"thought-provoking but ultimately speculative."* Others simply walked away without engaging, unwilling to consider any idea that blurred the lines between physics and philosophy.

That was fine.

As he collected his materials, a younger researcher approached him, a nervous but excited look on his face.

"I wanted to say," the researcher hesitated, *"I think you're onto something. I've been working on emergent field theories, and your results actually make me wonder if we're missing something*

fundamental about equilibrium. Maybe consciousness—well, maybe it's more than we think."

Kartik smiled. It was a small victory, but a victory nonetheless.

Not everyone in that room had to accept his ideas today.

Because he had planted a seed.

A thought, a doubt, a possibility—something that might grow over time, when the right evidence, the right experiment, or the right question arose.

And that was how real change began.

Kartik walked out of the auditorium not with certainty, but with conviction.

The journey was far from over.

In fact, it had only just begun.

A Personal Realization

Kartik sat in a quiet café near the symposium venue, staring at the untouched cup of coffee in front of him. The conversations, the challenges, the unexpected moments of support—all of it replayed in his mind. He had walked into that auditorium expecting resistance, but he had also hoped for curiosity.

And that was exactly what he had found—a divided audience, a room filled with both skeptics and open minds.

He had not walked away with universal validation, but he had not been dismissed outright either.

And that was not a failure.

It was proof of something far more important—ideas that challenge the status quo take time to take root.

He thought about the history of science. Every paradigm shift had been met with doubt.

- Quantum mechanics had been ridiculed before it reshaped physics.
- The idea of a curved space-time had been dismissed before it became essential to understanding gravity.
- Consciousness itself was once ignored by science—and even now, it remained one of the greatest unresolved mysteries.

What he had proposed today was no different.

It wasn't about proving that ancient knowledge was scientific. It wasn't about forcing modern physics to accept spiritual perspectives.

It was about expanding the conversation.

And in that, he had succeeded.

He took a deep breath, staring out of the window. The world outside looked the same—people walking, cars passing, conversations flowing. But for him, everything had changed.

This was not just a research topic anymore. This was a path—a journey few had dared to take.

And there was no turning back now.

He picked up his notebook, flipping to a fresh page. There was still so much more to explore.

And he was ready.

Part V — Toward A New Paradigm

Neither science nor scripture—just deeper questions

Chapter 9: The Search for a Unified Understanding

Kartik sat in his office, the glow of his laptop screen casting sharp shadows across the room. The symposium had ended, but its echoes still lingered in his mind. The divided reactions, the sharp criticisms, the unexpected support. He had walked into that lecture hall expecting resistance, but what he hadn't anticipated was the depth of the questions it had forced him to ask himself.

For months, he had been searching for a way to bridge science and ancient wisdom—to decode rather than dismiss. But now, he realized that this was never about proving one right over the other.

It was about uncovering something deeper—something neither had fully grasped.

The data from his research was real. The patterns in structured energy fields were undeniable. The stabilizing forces in equilibrium, the self-regulating interactions—they pointed to something fundamental, something modern physics had yet to quantify.

But could it be measured in the ways science demanded?

And more importantly—was he even asking the right questions?

Kartik leaned back, staring at the ceiling.

The journey was no longer about defending his findings.

It was about redefining the search for truth itself.

Revisiting the Fundamentals: Lessons from the Symposium

Kartik stared at his notes, flipping through the pages filled with equations, observations, and hastily scribbled thoughts from the past few months. The symposium had ended, but its impact lingered in his mind like an unresolved equation.

The reactions had been predictable yet revealing. Some had dismissed his findings outright, unwilling to engage with anything that blurred the lines between physics and philosophy. Others had been cautiously intrigued, willing to entertain the idea that knowledge might exist in forms beyond traditional scientific inquiry.

But what struck him the most was not the resistance—it was the questions his presentation had triggered.

It was never about convincing people that Vedic texts contained scientific truths.

It was about something much bigger: Had science been limiting itself by rejecting ideas simply because they came from non-conventional sources?

That was the real question.

He exhaled, running a hand through his hair. His mistake had been assuming this was a debate between two worldviews—science and ancient wisdom. But now, he saw it differently. This wasn't about proving one right over the other.

It was about refining the way science asked its questions.

For centuries, scientific discovery had been driven by expanding perspectives, by stepping beyond the known boundaries. The greatest breakthroughs had come from those who dared to question the assumptions everyone else accepted.

What if the role of ancient knowledge wasn't to replace science, but to challenge its blind spots?

What if the missing link between energy structures, equilibrium, and consciousness wasn't something that could be fully understood by equations alone?

Maybe science's greatest limitation was not in what it had discovered, but in what it had chosen to ignore.

Kartik leaned forward, his pen tapping against the desk.

The symposium had not been a victory or a failure.

It had been a catalyst.

And now, he knew where his work needed to go next. Not toward proving a theory, but toward expanding the very nature of scientific inquiry itself.

A New Model Emerges: Bridging Science and Consciousness

Kartik knew that to move forward, he needed more than isolated experiments and philosophical insights. He needed a structured framework—one that could be tested, refined, and challenged within the scientific community.

The symposium had proven that there was interest in his findings, even among skeptics. But if he wanted to shift the conversation from debate to discovery, he had to build a model—something measurable, something predictive.

He reached out to a select group of researchers—physicists, neuroscientists, and philosophers—those who had shown an openness to interdisciplinary approaches. Some were intrigued by the potential implications of his work, others were hesitant but curious. A few were outright skeptical but willing to test the validity of his claims through rigorous analysis.

Defining the Core Hypothesis

Over weeks of discussions, equations, and thought experiments, the team began to refine a core hypothesis:

- Consciousness is not merely an emergent property of the brain but an interactive component of structured energy fields.
- Self-regulating energy dynamics exist at fundamental levels, potentially influencing not just physical systems but also cognition and perception.
- The patterns Kartik had observed in structured energy equilibrium might be indicative of a deeper organizing principle—one that bridges the subjective and the objective.

If these principles were correct, they could redefine how science approached both physics and consciousness—not as separate phenomena, but as two expressions of the same underlying reality.

Formalizing the Mathematics

Kartik worked closely with Dr. Narayan Iyer, a mathematician specializing in nonlinear dynamics and self-organizing systems. Together, they began formalizing mathematical expressions that could describe self-regulating energy fields.

"This isn't just about structured fluctuations," Dr. Iyer noted as they worked through simulations. "It's about how these fluctuations interact. If a stabilizing factor exists, it must have an equation governing it—something akin to a feedback system that prevents pure entropy."

That was the missing piece.

They started constructing a dynamic feedback model, borrowing from control theory and emergent systems analysis. Kartik's experiments had shown that certain structured fields exhibited self-regulation, but now they needed to mathematically describe why.

The equations took shape, revealing a system that was not entirely random, yet not rigidly deterministic either. It hinted at an underlying balance—an unseen regulator, not unlike the ancient concept of Rta.

Neuroscience and the Consciousness Factor

At the same time, Kartik engaged with Dr. Aisha Patel, the cognitive scientist who had shown interest at the symposium.

"If you're suggesting that consciousness interacts with structured energy fields, then we need to determine whether perception influences these fluctuations in any measurable way," she pointed out.

Together, they designed an experiment to test whether the presence of an observer—or conscious intention—could subtly influence structured energy dynamics.

It was a bold move.

If they could show even a small effect, it would suggest that awareness was not just an internal process within the brain, but something capable of interacting with external systems.

The Emerging Model: Science, Energy, and Awareness

As weeks turned into months, the model began to take form.

It did not claim that consciousness was mystical—it suggested that structured equilibrium might be a missing component in understanding reality itself.

The principles they were now exploring could impact multiple fields:

- **Physics**: Understanding energy fields as self-organizing rather than purely entropic.
- **Neuroscience**: Investigating whether perception was tied to fundamental energy structures.
- **Philosophy**: Revisiting ancient ideas about consciousness and its role in shaping experience.

Kartik sat back, staring at the whiteboard covered in equations and diagrams.

For the first time, he saw a path forward.

This wasn't about validating ancient texts or forcing spirituality into science.

It was about asking the right questions—questions that modern science had yet to fully explore.

And now, he had a framework to begin answering them.

Facing the Gaps: The Limits of Both Science and Ancient Thought

Kartik leaned back in his chair, staring at the dense equations sprawled across his whiteboard. Something was missing.

Despite weeks of refining the theoretical model—describing self-regulating energy fields, equilibrium principles, and consciousness as an interactive component—there were still gaps that neither modern science nor ancient philosophy could fully address.

The Vedic descriptions of Rta and Brahman were elegant and intuitive, but they lacked quantifiable expressions. They spoke of order, balance, and interconnectedness, but they did not provide the mathematical rigor needed to fit within modern scientific frameworks.

On the other hand, modern physics had its own limitations. Quantum mechanics, relativity, and emergent systems theory had made extraordinary discoveries, yet they still struggled with fundamental paradoxes—the nature of time, the observer effect, and the role of consciousness in shaping reality.

Both perspectives offered pieces of the puzzle, but neither provided the full picture.

The Strengths and Weaknesses of Two Knowledge Systems

Kartik divided his notes into two sections—one for ancient insights, another for scientific frameworks.

- Ancient Philosophy provided a broad, experiential understanding of reality, speaking of consciousness, interconnectedness, and self-regulation. But it lacked precise, testable models.

- Modern Science had the mathematical rigor and empirical validation, but it avoided or dismissed ideas that could not yet be measured—particularly consciousness and self-regulating principles in energy fields.

This left an undeniable question—was the problem in the knowledge itself, or was it in the method of inquiry?

Had science limited itself by insisting that all truth must be measurable before it could be explored?

Had philosophy lost its ground by failing to evolve into testable hypotheses?

Kartik realized that true understanding required neither rejection nor blind acceptance—it required synthesis.

This was not a battle between worldviews—it was an attempt to bridge their strengths while leaving behind their respective weaknesses.

The Path Forward: A New Kind of Inquiry

Instead of proving one system over another, Kartik saw a different approach:

- Use scientific tools to test philosophical insights—rather than discarding ideas simply because they originated outside mainstream science.
- Expand the scientific method to explore consciousness as a fundamental force, not just an emergent property.
- Develop a framework where structured energy

equilibrium and cognitive perception could be tested together.

This wasn't about redefining science or proving philosophy right.

It was about evolving science to ask better questions.

Kartik picked up his notebook, writing one phrase at the top of the page:

"Where knowledge systems intersect, discovery begins."

He wasn't here to choose between two worlds.

He was here to build the bridge between them.

The Final Experiment: A Test of Consciousness and Structured Energy

Kartik stood in his lab, staring at the experimental setup before him. This was it.

Every previous test, every equation, every debate had led him here—to one final experiment that could either validate his hypothesis or dismantle everything he had been working toward.

If consciousness was not merely a byproduct of the brain but an interactive force within structured energy fields, there had to be a way to test it. Not through belief, not through philosophy—but through measurable, repeatable data.

Designing the Experiment

He worked meticulously, refining every detail to eliminate biases and ensure his findings would hold under scrutiny.

- **The Hypothesis:** If structured energy fields exhibit self-regulation, could an observer's intention or focus alter these stabilizing patterns?

- **The Setup:** A double-blind controlled study using precision measurement instruments to track fluctuations in structured energy fields.
- **The Variable:** Conscious focus. A select group of participants—some trained in deep meditation, others with no prior experience—would attempt to influence energy fluctuations without physical intervention.
- **The Control Measures:** Automated recording systems, randomized testing conditions, and external observers ensuring no human error or interference.

This wasn't about proving mind-over-matter. This was about testing whether consciousness—whatever it truly was—had a measurable impact on structured energy.

Executing the Test

The experiment began.

Participants entered the controlled chamber one by one, focusing on the structured energy field while sensitive detectors tracked fluctuations.

For hours, Kartik and his team monitored the results—scanning for patterns, deviations, any indication of an interaction.

The data streamed in.

Some fluctuations were within expected noise levels. But then, something changed.

A subtle but consistent shift.

Not large enough to be called proof, but too precise to be dismissed as random noise.

Over repeated trials, the pattern remained—the energy field exhibited slight but detectable changes when conscious focus was introduced.

The Aftermath: What Does It Mean?

Kartik stared at the data, his mind racing.

The results weren't definitive proof—but they were something.

They suggested that consciousness might interact with structured energy fields, even if the mechanism was still unknown.

It wasn't enough to rewrite physics overnight. But it was a crack in the rigid boundaries science had placed around the nature of consciousness.

This wasn't the end of his journey.

It was the beginning of a new frontier.

A question remained, lingering in the air like an unresolved equation:

What if consciousness wasn't just an effect of reality—but a force shaping it?

Kartik didn't have the full answer.

But now, he knew where to look next.

Beyond Science, Beyond Faith: An Unsettling Discovery

Kartik sat in his lab, staring at the data. The numbers flickered on the screen, confirming what he already knew—something had happened. The experiment had produced results that could not be dismissed as mere noise.

And yet, it was not enough.

The fluctuations in structured energy fields were measurable, consistent across trials, and yet they did not fit within any known model. They were neither strong enough to constitute definitive proof of consciousness interacting with physical reality, nor weak enough to be dismissed as random variation.

It was somewhere in between—a space science rarely liked to acknowledge.

A Truth That Refuses to Be Categorized

Kartik had expected two possible outcomes—either his hypothesis would be validated beyond doubt, or it would be disproven entirely.

But reality was rarely so simple.

What he had found was something real, but something that refused to conform neatly to the frameworks of either modern physics or ancient wisdom.

It wasn't faith, and it wasn't science as currently defined.

It was something else.

He thought back to the Vedic descriptions of Rta, the unseen order that sustained balance in the cosmos. Ancient thinkers had intuited that reality operated within an underlying structure, a principle of harmony that could not be seen but could be observed through its effects.

Now, his data hinted at the same possibility—but through entirely different means.

It was not about choosing between ancient wisdom and modern physics.

Both had seen parts of the truth, but neither held the complete picture.

Redefining the Way We Ask Questions

Kartik leaned back, rubbing his temples. Perhaps the real problem wasn't in the answers—but in the questions science had been asking all along.

Science had sought to measure reality as something external, separate from the observer. But if consciousness was fundamental, then perhaps reality was not something to be measured—but something to be understood from within.

Faith had often sought to accept truths without questioning them. But if structured energy fields responded to observation, then perhaps faith, too, had only glimpsed a fraction of something much deeper.

Neither science nor spirituality was wrong.

But both were incomplete.

The real challenge was not in proving one over the other—it was in expanding the way humans sought truth altogether.

Kartik exhaled, feeling the weight of realization settle over him.

This was bigger than a single experiment.

Bigger than him.

And he had only scratched the surface.

The Next Step: A New Way of Asking Questions

Kartik stood at the threshold of something far greater than he had anticipated. The experiment had not given him the final answer, but it had reshaped the way he saw the question.

For months, he had tried to reconcile two seemingly opposing views—science and ancient wisdom, equations and intuition, empirical proof and experiential knowledge. But the deeper he went, the more he realized that these were not opposing forces at all.

Science had always been about discovery, about refining models, discarding old assumptions, and venturing into the unknown. But somewhere along the way, it had begun policing the very boundaries it once sought to expand.

Spiritual traditions, on the other hand, had arrived at insights about consciousness, equilibrium, and interconnectedness long before science had the tools to measure them. Yet, without formal validation, those ideas had remained locked in the realm of belief rather than being explored as potential knowledge.

Kartik had spent much of his journey trying to decide which was right.

Now, he saw that this was never a choice between two opposing systems—but an invitation to rethink how humanity pursues truth.

Beyond the Known: Stepping into Uncharted Territory

His next steps would not be about proving ancient wisdom was correct, nor about forcing science to fit within rigid frameworks.

Instead, he would seek to expand how science approaches the unknown.

- He would refine the models he had built, searching for testable mechanisms that could describe the self-regulating patterns he had observed.

- He would collaborate with researchers outside traditional fields, bringing together physicists, neuroscientists, and cognitive scientists to explore the missing links between structured energy and perception.
- He would challenge the assumption that science must first measure before it can explore, advocating for new methodologies that allow inquiry beyond what current instruments can detect.

Kartik knew he was stepping beyond conventional boundaries.

The road ahead would not be easy. His work would face resistance, skepticism, and misinterpretation.

But if history had shown anything, it was that progress was made by those willing to question what others had deemed unquestionable.

He looked down at his notes, where a single sentence was scribbled in the margins:

"Truth is not a fixed destination—it is a path that must always be walked."

And he was ready to take the next step.

Chapter 10: The Unwritten Code

Kartik sat on a quiet bench outside the research institute, watching the world move around him. Conversations hummed in the background, the soft rustling of leaves carried in the wind, and the distant rhythm of a city continuing as if nothing had changed.

And yet, for him, everything had changed.

The symposium, the experiments, the debates—none had given him the final answer he had once sought. Instead, they had left him with something far more valuable: a new way of asking questions.

For months, he had tried to bridge two worlds—science and ancient wisdom—only to realize that they were never truly separate.

The search for truth was not about proving one system right over another, but about recognizing that knowledge is never complete.

His research had opened doors, but it had not provided closure. And maybe, it never would.

Because the greatest realization of all was this—truth is not something written, fixed, or final.

It is unwritten.

A pursuit that never ends, only evolves.

The Bigger Picture: Beyond Proving Ancient Knowledge

Kartik leaned back in his chair, staring at the blank page in his notebook. He had filled countless pages over the past few months—equations, observations, notes from the symposium, moments of doubt, flashes of realization. He had dissected and analyzed, tested and retested, debated and defended. And yet, now, as he reached what should have been the end of his journey, there was no final answer.

And for the first time, he realized that was exactly the point.

He had started this journey believing that he was searching for proof—proof that the Vedic descriptions of consciousness, energy, and balance aligned with modern physics. He had thought that if he could demonstrate the parallels with scientific rigor, he would have solved the puzzle—bridged the gap between science and ancient wisdom once and for all.

But now, sitting amidst his scattered notes, he saw how flawed that approach had been.

This was never about proving ancient knowledge through modern science.

It was about questioning how science itself approached fundamental truths.

The Limits of a Rigid Scientific Model

Science had evolved by breaking past its own limitations. Every major discovery had, at some point, been ridiculed, dismissed, or rejected because it did not fit within the existing framework of knowledge.

- Copernicus had faced rejection when he proposed that the Earth was not the center of the universe.
- Quantum mechanics had been dismissed as absurd until its mathematical foundations proved undeniable.
- Neuroscience still struggled to explain subjective experience—what it meant to *be* aware.

Kartik realized that the biggest limitation of science was not its methodology, but its unwillingness to consider knowledge that had not yet been quantified.

Ancient wisdom had taken a different approach. Instead of measurement, it relied on direct experience. Instead of equations, it spoke in principles.

Neither method was inherently wrong.

But both were incomplete.

The Flaw in 'Either-Or' Thinking

For too long, the debate had been framed as a choice between science and ancient wisdom.

- Either one accepted that science was the only valid way to explore reality, dismissing all other knowledge systems as unscientific.
- Or one placed blind faith in ancient texts, assuming they had all the answers even before modern validation.

Kartik saw now that this was a false choice.

Reality was not a battle between the measurable and the intuitive.

The real challenge was not to replace one system with another, but to expand our framework of understanding.

A Shift in Scientific Inquiry

What if science stopped limiting itself to what could already be measured?

What if it created new tools to explore what had only been intuited so far?

What if the scientific method evolved not by discarding ancient wisdom, but by refining its ability to test insights that had previously been inaccessible?

Kartik realized that the problem was never that ancient knowledge was unscientific.

The problem was that science had yet to develop the means to explore it.

And if history had taught anything, it was that the greatest advancements came not from certainty, but from the willingness to question.

A New Direction: Asking Better Questions

Kartik closed his notebook, but this time, it did not feel like the end of a journey.

It felt like the beginning of a new way of thinking.

He was no longer concerned with proving that ancient knowledge was correct—that had never been the real goal.

Instead, he was now asking:

- How can science develop better methods to explore

concepts like consciousness and structured energy equilibrium?

- How can ancient insights guide new lines of research instead of being discarded as philosophical speculation?
- What if the limitations of science were not in its discoveries, but in the questions it had never dared to ask?

For the first time, Kartik understood that truth is not fixed—it evolves.

Not through belief.

Not through rejection.

But through the courage to keep questioning.

And that, he realized, was the real purpose of his journey all along.

Science Without Boundaries: Rethinking Knowledge

Kartik sat in his office, the pages of his research spread across the table, interwoven with passages from ancient texts, mathematical equations, and notes from recent discussions with physicists and neuroscientists. As he sifted through them, a realization settled over him—knowledge was never meant to be bound by one framework, one culture, or one era.

Science had long prided itself on being universal, yet in practice, it had confined itself to a narrow methodology—one that valued measurement, reproducibility, and predictability above all else. That system had propelled humanity forward in remarkable ways, uncovering the forces of gravity, the nature of atoms, and the vast complexities of the universe.

But had it also restricted itself unnecessarily?

He thought back to the symposium. One of the strongest objections to his research had not been the data itself—but its associations. The moment he had introduced Vedic parallels, resistance hardened. Why? Because modern science had drawn an invisible line between what was considered knowledge and what was not.

The Arbitrary Boundaries of Knowledge

The more Kartik reflected, the more he realized that scientific inquiry had always been shaped by more than just reason—it had also been shaped by cultural and historical biases.

Ancient civilizations had all contributed to humanity's understanding of reality:

- Greek philosophers laid the groundwork for rational inquiry and logical deduction.
- Indian scholars developed nonlinear mathematics, the concept of zero, and deep cosmological theories.
- Chinese philosophers studied energy flow in ways that hinted at structured equilibrium systems.
- Indigenous traditions had long spoken of interconnectedness between life and the cosmos, a concept science was only now exploring through systems theory and ecological studies.

Yet, modern science had often dismissed knowledge that did not originate in its own historical narrative.

It had valued the European Renaissance over the contributions of the Islamic Golden Age, despite the fact that many classical ideas had been preserved and advanced by Arabic scholars.

It had ignored the mathematical sophistication of Indian and Mayan civilizations, treating them as curiosities rather than valid streams of knowledge.

And now, it hesitated to engage with ancient philosophical insights on consciousness and energy, fearing they would blur the lines between science and metaphysics.

But why should knowledge be confined by culture, time, or even the method through which it was first expressed?

The real question was not whether ancient insights were "scientific" in the modern sense—but whether they contained truths that modern science had yet to explore.

Breaking Free from Intellectual Borders

Kartik realized that science had reached a point where it needed to evolve—not by abandoning its rigor, but by expanding its scope.

The pursuit of knowledge should not be about choosing one methodology over another—but about integrating different perspectives to build a more complete picture of reality.

This did not mean that every ancient belief held scientific truth. It meant that dismissing ideas simply because they came from a non-modern framework was as unscientific as blind faith itself.

- If a civilization had observed patterns of equilibrium in nature and described them through metaphor, did that make the insight invalid?

- If a philosophical tradition had spoken of consciousness as fundamental, and modern physics was now questioning the role of the observer, was that a coincidence or a lost thread of inquiry?
- If history had shown that science evolves by questioning assumptions rather than protecting them, wasn't it time to challenge the assumption that knowledge must fit a specific cultural or methodological box?

Kartik put his pen down, exhaling.

Science was supposed to be about discovery.

But true discovery required the courage to look beyond familiar boundaries.

And maybe, just maybe, the next great breakthrough wouldn't come from where science had been looking—but from where it had refused to look.

Revisiting the Role of Consciousness: A Missing Variable?

Kartik stood at the edge of his research, staring at the one question science had been reluctant to confront head-on—consciousness.

For centuries, physics had reduced reality to particles, forces, and equations, while neuroscience had attempted to map thought and experience onto the complexity of the brain. And yet, after all this progress, one fundamental mystery remained:

What is consciousness, and does it play a role beyond subjective experience?

His experiment had shown subtle but measurable interactions between structured energy fields and observer awareness. It wasn't enough to claim that consciousness shaped reality in the way some interpretations of quantum mechanics suggested—but it was enough to suggest that the question could no longer be ignored.

If consciousness interacted with physical systems in even the smallest way, then science had been operating with an incomplete model of reality.

Physics: The Missing Variable in the Equation

Physics had been built on describing the external world—motion, forces, energy, space, and time. It had done so with unparalleled precision.

And yet, the deeper it probed, the more paradoxes emerged.

- Quantum mechanics had revealed that observation could influence measurement, yet physicists remained uncomfortable acknowledging why this happened.
- The search for a unified theory continued to fail because it could not reconcile how reality behaved at different scales—from subatomic particles to galaxies.
- Dark energy and dark matter remained unsolved mysteries, suggesting that science was missing something fundamental about the nature of existence.

If consciousness was not just an illusion produced by the brain but an active part of reality, then perhaps the equations of physics needed to evolve.

Not by discarding material science, but by recognizing that something was missing in its framework.

Neuroscience: Is the Brain a Generator or a Receiver?

For decades, neuroscience had approached consciousness as a byproduct of brain function—the result of neurons firing, electrical impulses flowing, and chemical reactions shaping cognition.

But what if the brain wasn't generating consciousness—but interacting with it?

The idea was radical, yet not entirely new.

- Some cognitive scientists had begun exploring consciousness as a field-like phenomenon, one that extended beyond individual neurons.
- The study of near-death experiences and anomalous cognition hinted at a reality that could not be explained solely through brain chemistry.
- Meditation research had already suggested that awareness could alter neural activity, raising the question—was consciousness local or non-local?

If consciousness interacted with structured energy fields, then neuroscience, too, would need to rethink its assumptions.

Philosophy: Consciousness as the Underlying Reality?

In Vedic thought, consciousness had always been considered fundamental—not something emerging from the body, but something that pervaded existence itself.

For centuries, philosophy and science had drifted apart, with the former considered speculative and the latter empirical.

But what if the division had been premature?

If modern physics was struggling to explain the observer effect, wasn't it time to reconsider the ancient claim that perception played a role in shaping experience?

If structured equilibrium existed at fundamental levels of energy, wasn't it worth revisiting the idea that consciousness itself might be an organizing principle?

This wasn't about returning to mysticism.

It was about recognizing that philosophy had been asking questions long before science had the tools to answer them.

Perhaps, the next breakthrough in understanding consciousness wouldn't come from physics or neuroscience alone—but from bridging both with the forgotten insights of philosophical traditions.

A New Paradigm: Reality Beyond the Material

Kartik exhaled, feeling the magnitude of what lay ahead.

He wasn't advocating for abandoning material science.

He wasn't suggesting replacing physics with philosophy.

But he was beginning to see that reality might not be purely materialistic.

Consciousness was not separate from existence.

It was time for science to stop treating it as an anomaly and start considering it as a variable—one that might just redefine how we understand the universe.

And that, he realized, was a question worth dedicating his life to answering.

A Legacy of Questions, Not Answers

Kartik stared at the final page of his research notes, but instead of a conclusion, all he saw was a beginning.

For months, he had searched for a definitive answer—a way to bridge science and consciousness, equations and experience, modern physics and ancient wisdom. But now, he understood that his journey had never been about finding a final truth.

Instead, it had led him to better questions.

Science had always evolved by expanding its boundaries, challenging its assumptions, and refining its methods. But was it ready to ask the questions that it had long ignored?

Can science expand beyond measurement-based inquiry?

- If reality contains elements that are not purely physical, does science need new tools of investigation beyond material measurement?
- Could experience itself—the first-person perspective of consciousness—become a valid part of scientific exploration?

If consciousness interacts with structured energy fields, how can this be modeled?

- His experiment had hinted at an interaction, but the mechanisms remained unknown.
- Was it a feedback system? A non-local effect? A missing force that science had yet to define?

What new experimental methods can be developed to test these ideas further?

- Could physics develop sensitive enough instruments to measure consciousness as a field?
- Could neuroscience shift from studying the brain alone to investigating the brain's interaction with external energy structures?

These were no longer just his questions.

They belonged to anyone willing to challenge the limits of knowledge.

Kartik had started this journey believing he was walking between two opposing worlds—science and philosophy, physics and consciousness.

Now, he realized there was no boundary at all.

Knowledge was not a fixed system but a moving horizon, constantly unfolding.

And he had no desire to close the book on it.

Instead, he would keep questioning.

Because truth was not something found—it was something pursued.

The Unwritten Code: The Endless Pursuit of Truth

Kartik stood by his office window, watching the city lights flicker in the distance. The world outside moved as it always had—unconcerned with the questions that consumed him. The symposium had ended. His research had reached a milestone. And yet, he felt no sense of completion.

Because truth, he now realized, was never meant to be a fixed destination.

He had spent months trying to bridge science and philosophy, to reconcile the material with the immaterial, to find a singular answer that could unify consciousness, structured energy, and the nature of existence.

But now, he understood that no single framework—scientific, philosophical, or otherwise—could fully encapsulate reality.

Every discovery was just a step toward a deeper question.

Every theory was a temporary construct, waiting to be refined or replaced.

This was the unwritten code of knowledge.

Not a hidden formula, not a final revelation—but the willingness to keep questioning.

Beyond Proof: The Courage to Explore

Kartik thought back to the great minds who had shaped science—Einstein, Bohr, Heisenberg, Hawking. None of them had settled for what was already known.

- Einstein had redefined space and time.
- Bohr had shattered classical physics with quantum uncertainty.
- Hawking had pushed the limits of what could be understood about black holes.

They hadn't been afraid to challenge the boundaries of human understanding.

And now, he realized, he was standing at his own threshold.

Not to prove.
Not to convince.
But to explore.

The Next Chapter: Asking the Right Questions

Kartik turned back to his desk, where his notebooks lay open, pages filled with equations, sketches, and scattered thoughts.

His research was far from over.

If anything, it was just beginning.

The real task ahead was not about validation—but expansion.

- How does consciousness interact with structured energy fields?
- Can science create new methodologies to explore non-material aspects of reality?
- What happens when we stop rejecting knowledge from different traditions and start integrating them into a unified pursuit of truth?

Kartik picked up his pen and wrote in the margins of his notebook:

"Truth is not something we arrive at—it is something we must constantly seek."

There would be more debates. More challenges. More resistance.

But he was ready.

Because truth does not belong to one era, one method, or one mind.

It belongs to those willing to pursue it—without fear, without boundaries, without end.

Epilogue

The Scientist Who Chose to Ask

Kartik sat in his study, the soft hum of the city outside his window a quiet reminder that the world moved forward, indifferent to the questions that consumed him. His desk was covered with notes, some filled with equations, others with philosophical reflections—evidence of a journey that had blurred the boundaries between science and ancient wisdom, between the measurable and the ineffable.

For months, he had sought answers—a way to bridge structured energy fields, consciousness, and the fundamental nature of reality. He had expected to arrive at a conclusion, a final understanding. But now, as he reached the end of this chapter in his research, he realized something profound:

There are no final answers.

The greatest pursuit of knowledge was never about proving or disproving—but about continuing to ask the questions that others had stopped asking.

The Future of Inquiry

Kartik's research had opened doors he had not anticipated. The experiment had shown subtle but measurable interactions between structured energy fields and conscious observation. It wasn't definitive—it wasn't the breakthrough that would rewrite physics overnight—but it was something.

And something was enough.

Because now, the challenge was no longer whether science could measure consciousness. The challenge was whether science could redefine itself to explore it.

- Could physics accept that the universe might not be purely material?
- Could neuroscience reconsider that awareness might not be just an illusion of the brain?
- Could scientific inquiry evolve beyond the constraints it had placed upon itself?

These were not just questions for him.

They were questions for the next generation of thinkers, scientists, and seekers—those who were willing to step beyond the known and venture into the unexplored.

The Unfinished Code

For the first time, Kartik no longer felt the pressure to prove.

His search had led him not to a hidden formula or a lost truth, but to the realization that truth itself is never fully written.

It is an unfinished code—one that continues to evolve with every question we dare to ask.

He picked up his notebook and turned to a blank page.

For years, he had filled pages with calculations, theories, and interpretations. But now, he stared at the empty space before him—not as an ending, but as an invitation.

There was more to discover.

More to challenge.

More to explore.

Because truth does not belong to the past, nor is it confined to the present.

It belongs to those who refuse to stop searching.

Kartik smiled to himself, closed his notebook, and turned off the light.

Tomorrow, the search would begin again.

Appendices

"The code is not in any scripture, formula, or lab—it's in the courage to look again."

Appendix A — Key Scientific Concepts

Foundational Ideas for Understanding the Science Behind Ancient Knowledge

This appendix provides clear, reader-friendly explanations of the scientific principles referenced throughout the book. These summaries are not exhaustive; rather, they serve as conceptual anchors that help bridge Kartik's scientific exploration with ancient philosophical insights.

1. Quantum Mechanics

Quantum mechanics is the branch of physics that studies reality at the smallest scales—particles, waves, and the strange behavior of the subatomic world. Its discoveries have challenged classical intuitions for over a century.

Key Ideas

- **Quantum Superposition** A particle can exist in multiple possible states simultaneously until it is observed. This principle fundamentally challenges the idea that reality has fixed properties independent of observation.
- **The Observer Effect** Measuring a quantum system

influences its behavior. The act of observation collapses multiple possibilities into a single outcome. This enigmatic effect lies at the heart of modern debates on the role of consciousness in physics.

- **Quantum Entanglement** Two particles become linked such that the state of one instantly affects the state of the other, regardless of distance. Einstein dismissed it as "spooky action at a distance," but experiments confirm its reality.

Why it matters for this book Quantum mechanics is the first scientific domain that requires an observer—and this opens a profound conversation about who or what is watching.

2. Thermodynamics & Entropy

Thermodynamics studies the flow of energy and heat. Entropy, one of its central principles, describes the degree of disorder in a system.

Key Ideas

- **The Second Law of Thermodynamics** Entropy tends to increase. Systems naturally move from order to disorder unless an external force stabilizes them.
- **Local Order vs. Universal Entropy** Even though the universe trends toward chaos, pockets of order still arise—atoms form stable structures, planets maintain orbits, life organizes itself.

Why it matters Vedic Rta—the principle of cosmic order—conceptually resonates with the balancing act between entropy and local order. Kartik's research into self-regulating energy fields stems from this very paradox.

3. Structured Energy Fields

In physics, fields are invisible frameworks that govern how forces operate—gravity, electromagnetism, and nuclear interactions all function through fields.

Key Ideas

- **Energy Field** A continuous region of space where each point has a measurable value of a physical quantity (like electric or gravitational potential).
- **Structured Equilibrium** Some energy fields do not merely dissipate—they self-stabilize, showing patterns of balance that are not random. Kartik's experiments explore whether consciousness can influence these structures.

Why it matters This idea is central to bridging ancient descriptions of cosmic order with scientific models of energy behavior.

4. Feedback Systems

Feedback loops are fundamental in biology, physics, technology, and cognitive sciences. They describe systems where outputs feed back into inputs, stabilizing or amplifying changes.

Types of Feedback

- **Negative Feedback** Stabilizes a system (e.g., homeostasis in human body temperature).

- **Positive Feedback** Amplifies effects (e.g., chain reactions, exponential growth).

Why feedback matters Kartik's structured energy model relies heavily on negative feedback—mechanisms that allow systems to "self-correct," something ancient philosophy intuitively described as harmony or balance.

5. Neuroscience & Consciousness

Neuroscience studies the brain; consciousness studies the subjective experience—the "inner world" no instrument can directly measure.

Key Ideas

- **Neural Correlates of Consciousness (NCC)** Brain regions linked to awareness. Modern neuroscience has identified activity patterns associated with perception but still cannot explain how the brain creates experience.
- **The Hard Problem of Consciousness** Coined by David Chalmers, it asks: How does subjective experience arise from physical processes? Science can track the electrical signals but cannot explain why you feel.
- **Integrated Information Theory (IIT)** Suggests consciousness is the result of how information integrates in a system—not limited to human brains.

Why it matters This is where science meets philosophy: Is consciousness produced by the brain, or does the brain receive or interact with it?

6. Systems Thinking in Science

Systems thinking views phenomena not as isolated parts but as interconnected wholes.

Key Ideas

- **Holism** The whole is greater than the sum of its parts. For example, life cannot be explained by analyzing molecules alone; interaction matters.
- **Emergent Properties** Some behaviors emerge only when parts interact (e.g., consciousness emerging from brain complexity—or perhaps interacting with external fields).
- **Complex Adaptive Systems** Systems that evolve, self-correct, and adapt—ecosystems, economies, organisms.

Why it matters Both Vedic cosmology and modern physics hint at a universe where relationships matter more than isolated components. This appendix provides the conceptual bridge.

Final Note on Appendix A

These scientific concepts are not here to prove ancient knowledge or reinterpret physics through a mystical lens. They exist to show that modern science and ancient insight often meet at unexpected intersections—and those intersections are where new understanding begins.

Appendix B — Vedic Concepts

& Their Scientific Correlations

A Comparative Framework Bridging Ancient Insight and Modern Scientific Thought

This appendix outlines key philosophical ideas from Vedic literature and aligns them with corresponding concepts in contemporary science—not as direct equivalents, but as conceptual parallels. The aim is not to force a match but to show where ancient experience and modern experimentation hint at similar truths.

1. Rta — The Principle of Cosmic Order

Vedic View Rta is the foundational law of natural and moral order in the universe. It governs the cycles of seasons, the motion of planets, the rhythm of life, and even ethical conduct.

Scientific Correlation Comparable to the concept of universal laws of physics—gravitational laws, conservation of energy, and thermodynamic equilibrium. It also resonates with entropy management, where order is preserved within pockets of increasing disorder.

2. Atman — The Inner Self or Observer

Vedic View Atman is the individual consciousness, a reflection of the universal Self (Brahman). It is not limited to mind or body but is the eternal, witnessing presence.

Scientific Correlation Aligns with the unresolved "hard problem of consciousness" in neuroscience and cognitive science. Echoes the quantum observer effect, where consciousness appears to influence outcome in a probabilistic system.

3. Brahman — The Universal Consciousness or Supreme Reality

Vedic View Brahman is the unchanging, formless source of all that exists. Everything arises from it and returns to it. It pervades all creation.

Scientific Correlation Parallels the idea of a unified field in physics, or a structured energy medium in quantum vacuum theories. Also aligns with recent speculation on a conscious universe in integrated information theory (IIT) and panpsychism.

4. Maya — Illusion or Perceived Reality

Vedic View Maya is not falsehood, but the temporary appearance of form and separateness—an illusion of fragmentation in an inherently unified reality.

Scientific Correlation Reflects simulation theories, quantum uncertainty, and philosophical interpretations of reality as emergent rather than fundamental. Echoes how perception in neuroscience constructs a version of the world.

5. Yajna — Sacrifice or Cyclical Exchange

Vedic View Yajna is the symbolic and practical process of offering and transformation—sacrifices made to maintain balance and harmony in nature.

Scientific Correlation Resonates with energy exchange, feedback loops, and thermodynamic cycles. Also mirrors ecological balance and systems theory, where equilibrium is maintained through dynamic input and output.

6. Prana — Vital Force or Life Energy

Vedic View Prana is the dynamic life force that flows through all living beings. It is responsible for movement, breath, vitality, and sensation.

Scientific Correlation Comparable to bioelectrical energy, nervous system activity, and the role of field interactions in living systems. While not equivalent to any one scientific term, it overlaps with electromagnetic regulation, homeostasis, and vibrational coherence.

7. Purusha & Prakriti — Consciousness and Nature

Vedic View Purusha is the pure consciousness (male principle), while Prakriti is the dynamic nature (female principle). All experience arises from the interplay between these two.

Scientific Correlation Suggests a dualistic view similar to observer vs. observed. Echoes subjective experience (consciousness) interacting with objective reality (matter) in quantum observation and measurement.

8. Kala — Time

Vedic View Kala is not just linear time, but cyclical and recursive. It governs birth, decay, and rebirth—reality unfolds in patterns, not straight lines.

Scientific Correlation Correlates with entropy, causality, and the arrow of time in physics. Also echoes time symmetry in quantum theory and recurrence patterns in cosmology and wave theory.

9. Nada — Cosmic Vibration or Sound

Vedic View Nada is the primordial vibration—the first expression of Brahman. The universe emerges from sound (Om) as frequency and form.

Scientific Correlation Relates to vibrational modes, wave-particle duality, resonance in quantum mechanics, and field oscillations. Some physicists theorize that reality itself may be formed of oscillating energy patterns.

10. Tattvas — Fundamental Elements or Principles

Vedic View Tattvas are the building blocks of existence, starting from subtle (space, mind) to gross (earth, water, fire).

Scientific Correlation Comparable to standard model particles, fields, forces, and dimensional hierarchies. Also aligns with the idea of emergent layers—from quantum fields to complex systems.

Closing Note on Appendix B

These correlations are not meant to equate spiritual wisdom with scientific fact. Instead, they point to a deeper convergence—that ancient seers and modern scientists may have been exploring different layers of the same reality, through vastly different tools: experience vs. experimentation.

By decoding these concepts without distortion, we can build a shared framework where science does not replace spirituality, and spirituality does not undermine science, but both evolve together in pursuit of truth.

Appendix C — Experimental Models & Hypotheses

A Cross-Disciplinary Exploration Grounded in Structured Energy and Consciousness Interactions

This appendix outlines the experimental thinking developed throughout the book, summarizing Kartik's approach and the broader implications for science. These models are designed not as final answers, but as hypotheses under refinement—starting points for future empirical exploration.

1. Structured Energy Equilibrium Model (SEEM)

Hypothesis All physical systems self-regulate through internal ripple-based energy feedback loops—what ancient texts referred to as "cosmic balance" (Rta).

Experimental Basis

- Controlled field simulations using phase-locked oscillators
- Resonance stability measurements in dynamic vs. static

environments

Key Metrics

- Energy dissipation patterns
- Self-restorative field behavior
- Fractal boundary stabilization

Modern Parallel

- Quantum field dynamics
- Non-linear thermodynamic systems
- Chaos and attractor models

2. Observer-Field Interaction Hypothesis

Hypothesis Conscious awareness can interact with structured energy fields in subtle but measurable ways.

Experimental Basis

- Blind interaction trials using intention-based field modulation
- EEG and EMF co-registrations during focused awareness experiments

Key Metrics

- Deviation in EM field coherence under focused vs. unfocused observation
- Local entropy shift compared to control baseline

Modern Parallel

- Quantum observer effect
- Noetic science trials
- Biocentric cosmology

3. Consciousness as a Non-Local Field Hypothesis

Hypothesis Consciousness is not generated by the brain but interfaces with the body through a distributed ripple field—similar to Vedic Atman-Brahman relationship.

Experimental Basis

- Near-death EEG coherence studies
- Remote synchronization of intention and physiological changes across distance

Key Metrics

- Anomalous signal correlation without known physical link
- Neuro-physiological baseline shift in recipient group

Modern Parallel

- Integrated Information Theory (IIT)
- Non-local entanglement models
- Rupert Sheldrake's morphic resonance studies (controversial but conceptually similar)

4. Vedic Concept-Inspired Modeling of Energy Flow

Hypothesis Energy does not behave randomly but follows structured flow paths that resemble Vedic metaphors like Yajna (cyclical offering) and Prana (directed force).

Experimental Basis

- Closed-loop energy feedback systems using phase-dependent logic gates
- Pattern mapping of energy distribution in cellular automata systems

Key Metrics

- Recurrent self-patterning
- Dynamic equilibrium after external perturbation
- Ripple-coherence propagation

Modern Parallel

- Self-organizing systems
- Emergent behavior in networked fields
- Biofield regulation in living systems

5. Integration Framework for Ancient Metaphor and Scientific Logic

Objective Not to validate spiritual beliefs through science, but to extract metaphorical frameworks that may inspire testable models in physics and consciousness studies.

Approach

- Symbolic deconstruction of Vedic verses
- Mapping of metaphors to scientific terminology
- Rebuilding models based on experimental possibility, not dogma

Example Mapping

- *"As is the atom, so is the cosmos"* → Fractal scaling / holographic models
- *"That which moves and moves not..."* → Frame of reference in relativity
- *"Prana flows in channels"* → Neural/electrical/magnetic pathway networks

General Notes for Researchers

- Replication is critical. Kartik's models are early-stage frameworks meant for rigorous validation.
- Bias must be eliminated. Interpretive frameworks should not overreach metaphysical claims.
- Interdisciplinary teams are essential. Philosophers, physicists, neuroscientists, and linguists must collaborate.

Final Reflection

These experimental frameworks do not claim superiority over traditional science or ancient wisdom—but suggest a third path: a convergent evolution of thought where insights from both worlds inspire new forms of inquiry.

The models presented here are not endpoints, but doorways.

Truth is not a destination—it's a resonance.

Appendix D — Quote Bank & Reflective Aphorisms

For Reflection, Research, and Reframing the Known

This appendix offers distilled insights—aphorisms, reflections, and powerful one-liners—from Kartik's journey and the thematic fabric of this work. These quotes can serve as starting points for contemplation, debate, research, or writing.

On Science and Ancient Wisdom

- "Science measures. Ancient wisdom observes. But truth transcends both."
- "Perhaps the ancients weren't trying to describe equations—they were describing experiences too consistent to be ignored."
- "Vedic thought didn't need a lab. It had the universe—and consciousness—as the instrument."
- "If a truth emerges in metaphor, it is not less true—only waiting for another method to grasp it."

On Consciousness and Perception

- "If the observer collapses the wave, then who observes the observer?"
- "Consciousness may not arise from the brain. It may simply wear the brain like a tool."
- "What you see is filtered through what you believe. This is not mysticism—it's cognitive physics."
- "Experience may be the missing metric of reality."

On Structured Energy and Balance

- "Order is not the opposite of entropy—it's the rhythm that lives between chaos and collapse."
- "Structured energy fields may be nature's way of thinking without neurons."
- "What Rta described poetically, physics now glimpses through equations: the dance of balance."
- "Energy is never random. It listens to patterns we have not yet learned to read."

On Doubt, Discovery, and Courage

- "The greatest breakthroughs do not begin with 'Eureka'—they begin with 'What if?'"
- "Skepticism without curiosity becomes cynicism. Belief without questions becomes blindness."
- "A closed mind fears what cannot be named. An open one learns to name what it cannot yet prove."
- "It is not irrational to ask. It is irrational to never ask simply because it doesn't fit a model."

On Kartik's Journey

- "He didn't want to believe—he just couldn't ignore the evidence anymore."
- "Kartik never crossed over to faith. He simply walked far enough that both science and spirituality were behind him."
- "He wasn't searching for God. He was searching for consistency—and found mystery along the way."
- "In seeking equations, he found metaphors. In decoding metaphors, he discovered equations."

On The Unwritten Code

- "The code is not in any scripture, formula, or lab—it's in the courage to look again."
- "The universe may not be asking to be understood. It may only be asking to be engaged with honesty."
- "Truth is not a system—it's a signal. Every framework receives only part of it."
- "The real frontier isn't outer space. It's the space between what we measure and what we experience."

Notes & References

This section provides references to the scientific studies, philosophical texts, and experimental research that have informed and supported the discussions in this book. It also includes my own published work, contributing to the exploration of consciousness, structured energy equilibrium, and the intersection of science and ancient wisdom.

Scientific References

Physics & Quantum Mechanics

- Heisenberg, W. *Physics and Philosophy: The Revolution in Modern Science*. Harper & Row, 1958.
- Bohm, D. *Wholeness and the Implicate Order*. Routledge, 1980.
- Penrose, R. *The Emperor's New Mind: Concerning Computers, Minds, and the Laws of Physics*. Oxford University Press, 1989.
- Wheeler, J.A. *Law Without Law: Quantum Theory and Reality*, in *Quantum Theory and Measurement*. Princeton

University Press, 1983.

- Chavan, S *The Blueprint of the Universe: The Theory That Could Change Everything We Know About Reality* 2024

Neuroscience & Consciousness Studies

- Koch, C. *The Feeling of Life Itself: Why Consciousness Is Widespread but Can't Be Computed.* MIT Press, 2019.
- Chalmers, D.J. *The Conscious Mind: In Search of a Fundamental Theory*. Oxford University Press, 1996.
- Tononi, G. *Integrated Information Theory of Consciousness. Nature Reviews Neuroscience*, 2016.

Philosophy & Ancient Knowledge

- Upanishads. *Brihadaranyaka Upanishad, Chandogya Upanishad, Mandukya Upanishad* (Various Translations).
- Dandekar, R.N. *Vedic Cosmology and Modern Science. Journal of Indian Philosophy*, 1975.
- Capra, F. *The Tao of Physics: An Exploration of the Parallels Between Modern Physics and Eastern Mysticism.* Shambhala Publications, 1975.
- Chavan, S *Energy in Action: The Soul Demystified*, 2024
- Chavan, S *The Active Mind: Functional Manifestation of the Soul,* 2024
- Chavan, S *Energy in Connection: Consciousness Demystified, 2024*

Interdisciplinary & Emerging Theories

- Hameroff, S., & Penrose, R. *Orchestrated Objective Reduction of Quantum States in Microtubules: A Model for Consciousness. Physics of Life Reviews*, 2014.
- Kauffman, S. *At Home in the Universe: The Search for the Laws of Self-Organization and Complexity.* Oxford University Press, 1995.
- Pribram, K. *The Holonomic Brain Theory in Neuroscience. Journal of Theoretical Biology*, 1991.

Additional Readings for Further Exploration

For readers interested in deeper exploration of the topics discussed in this book, the following books and papers provide valuable insights:

- Stapp, H. *Mind, Matter, and Quantum Mechanics.* Springer, 2004.
- Barrow, J.D. *The Constants of Nature: The Numbers That Encode the Deepest Secrets of the Universe.* Vintage, 2002.
- Nagel, T. *Mind and Cosmos: Why the Materialist Neo-Darwinian Conception of Nature Is Almost Certainly False.* Oxford University Press, 2012.

This collection of references provides both scientific rigor and philosophical depth, allowing readers to explore the broader context of structured energy fields, consciousness, and the evolution of scientific inquiry.

The Road Ahead

As this journey comes to a close, one truth stands clear—this is not the end, but the beginning of a much larger inquiry.

The exploration of consciousness, structured energy, and the fundamental nature of reality has opened doors that science has yet to walk through. While the research presented here has provided evidence of patterns, interactions, and unresolved gaps in current models, it has also raised more questions than answers.

And that is how it should be.

Knowledge is not static—it evolves. Every great shift in human understanding has begun not with absolute proof, but with the courage to ask the right questions.

Unanswered Questions & Future Research Directions

The work ahead must continue to explore the following fundamental questions:

Can science expand beyond measurement-based inquiry?

- If structured energy fields and consciousness interact, what new methodologies can be developed to study them?
- How do we refine scientific tools to detect subtle, non-

material influences in physical systems?

What is the nature of consciousness, and how does it interact with reality?

- Is consciousness an emergent property of the brain, or does it exist as a field-like phenomenon that interacts with matter and energy?
- Can consciousness be integrated into the fundamental equations of physics, just as space, time, and energy have been?

How do structured energy fields contribute to equilibrium in the universe?

- Are these patterns simply the result of natural laws, or do they suggest an underlying order that physics has yet to define?
- Could this be the missing link in unifying quantum mechanics and relativity?

How can ancient knowledge contribute to modern scientific frameworks?

- Instead of discarding ancient perspectives as speculative, how can we use them as a starting point for testable hypotheses?
- What insights might Vedic and other ancient traditions hold that align with the yet-to-be-explored dimensions of physics and consciousness?

A Call for Open Inquiry

The challenge of the future is not just to refine existing theories, but to expand the very way we ask scientific questions.

If history has shown anything, it is that science advances when it dares to challenge its own assumptions.

The same spirit that once led to the discovery of quantum mechanics, relativity, and neural plasticity must now be applied to the exploration of consciousness and its role in the universe.

This requires a new kind of scientific openness—one that does not dismiss ideas because they originate from philosophy, history, or ancient traditions, but instead tests them with the same rigor as any other scientific hypothesis.

The road ahead is not about proving one worldview over another.

It is about recognizing that truth is always evolving—and that to find it, we must be willing to explore the unknown.

This is not just my journey.

It is an open invitation to every thinker, scientist, philosopher, and seeker who believes that the pursuit of knowledge is limitless.

The real question now is—who will take the next step?

Acknowledgments

This book is the result of a journey that extends far beyond my own research and reflections. It is shaped by the minds, challenges, and inspirations of many individuals who have, in different ways, contributed to this exploration.

First and foremost, I extend my gratitude to the scientists, physicists, and researchers whose groundbreaking work has laid the foundation for our understanding of reality. Their relentless pursuit of truth, their ability to question the very fabric of existence, and their willingness to challenge the unknown have inspired me to push beyond conventional boundaries.

To the philosophers, scholars, and historians of ancient wisdom, whose insights into consciousness, energy, and the nature of reality have stood the test of time. While science continues to refine its models, their intuitive grasp of interconnectedness and balance has provided invaluable perspectives that remain as relevant today as they were centuries ago.

To my mentors and colleagues, who engaged in thought-provoking discussions, challenged my assumptions, and forced me to refine my arguments. Their skepticism was not a barrier but a necessary force that sharpened the clarity of this work.

To the skeptics—both those I encountered in academic circles and those who debated my ideas—thank you. Your resistance, your demands for proof, and your critical perspectives have been instrumental in shaping this research with greater depth and rigor.

To my friends and family, who have patiently endured my endless discussions on consciousness, physics, and philosophy. Your support, even when you did not fully agree with my ideas, gave me the strength to continue.

Finally, to the readers and seekers, those who refuse to accept easy answers, who continue to question, and who are unafraid to challenge what they know—this book is for you. The pursuit of truth is never about arriving at a final answer, but about expanding the way we ask questions.

May we all continue searching, exploring, and pushing the boundaries of knowledge—together.

About the Author

Sandeep Chavan is an independent researcher, author, and thinker exploring the intersections of science, consciousness, and ancient wisdom. With a background in engineering, industrial sciences, and philosophical inquiry, he has spent years examining the nature of structured energy, self-regulating systems, and the role of consciousness in shaping reality.

His work is driven by a fundamental question: Has modern science overlooked key insights that ancient civilizations intuitively understood? This inquiry has led him to explore fields ranging from quantum mechanics and neuroscience to Vedic philosophy and emergent systems.

Motivation Behind This Book

"The Unwritten Code: Exploring the Science Behind Ancient Knowledge" was born out of a desire to bridge scientific inquiry with forgotten wisdom. The book challenges the conventional limitations of knowledge and asks whether science can evolve to explore the very mysteries it has long ignored—the nature of consciousness, the unseen order in the universe, and the fundamental structure of reality.

Through rigorous analysis, experimentation, and open-ended questioning, Sandeep Chavan presents a vision where science does not reject the past, but instead expands to explore the deeper truths that remain just beyond its reach.

Future Projects & Research Directions

His ongoing work continues to explore:

- The role of structured energy fields in consciousness.
- New experimental methodologies to test non-material influences on physical systems.
- Interdisciplinary models that integrate physics, neuroscience, and ancient philosophical insights.

Stay Connected

Readers interested in following Sandeep Chavan's work, future publications, and research updates can connect via:

Books & Publications – Amazon, Apple Books, Barnes & Noble, Kobo, Everand, Smashwords, Tolino, OverDrive, cloudLibrary, Baker & Taylor, BorrowBox, Hoopla, Vivlio, Palace Marketplace, Odilo, Gardners, Fable etc

Research Papers – Zenodo[1]

For discussions, collaborations, or inquiries, feel free to reach out.

The search for truth is never-ending—and the greatest discoveries lie in the questions we have yet to ask.

1. https://doi.org/10.5281/zenodo.14793162

Disclaimer

The content of *The Unwritten Code: Exploring the Science Behind Ancient Knowledge* is intended for informational and exploratory purposes only. This book does not claim to provide definitive scientific proof of ancient philosophical concepts, nor does it advocate for replacing established scientific methodologies with spiritual or metaphysical beliefs.

The discussions in this book are based on scientific theories, experimental observations, historical texts, and philosophical perspectives. While every effort has been made to ensure the accuracy of the information presented, readers are encouraged to approach the material with critical thinking and an open mind.

This book is not a scientific textbook, religious doctrine, or a substitute for formal academic research. It is an exploration of ideas, inviting interdisciplinary dialogue between physics, consciousness studies, and ancient wisdom traditions.

The views expressed in this book are those of the author and do not necessarily reflect the opinions of any institution, organization, or research body. Readers are advised to conduct their own research and consult experts in relevant fields before forming conclusions or applying any of the ideas discussed herein.

Science is an evolving process, and the pursuit of truth is an ongoing journey. This book serves as an invitation to question, explore, and expand the boundaries of inquiry—without rigid adherence to any single perspective.

The author and publisher assume no responsibility for interpretations or applications of the material presented in this book.

Don't miss out!

Visit the website below and you can sign up to receive emails whenever Sandeep Chavan publishes a new book. There's no charge and no obligation.

https://books2read.com/r/B-A-EPGPC-VUEKI

BOOKS 2 READ

Connecting independent readers to independent writers.

Did you love *The Unwritten Code*? Then you should read *Dhritarashtra: The Blindness Within*[1] by Sandeep Chavan!

[2]

Dhritarashtra: The Blindness Within is not a retelling of the Mahabharata — it is a decoding of human blindness.It asks one haunting question: **What happens when power sees everything except truth?**

Through the lens of **Ripple Field Philosophy** — a unifying framework connecting energy, ethics, and awareness — **Sandeep J. Chavan** reinterprets the archetype of *Dhritarashtra*, the blind king, as a timeless pattern of human consciousness. This blindness is not

1. https://books2read.com/u/4EX70O
2. https://books2read.com/u/4EX70O

physical; it is moral and psychological — the refusal to see what awareness reveals.

The book unfolds in four interconnected parts:

Part I – The Archetype of Blindness: explores Dhritarashtra as an energy field — the origin of distortion where perception detaches from awareness.**Part II – The Collapse of Consequence:** reveals how moral decay begins with emotional blindness and spreads through power, loyalty, and control.**Part III – The Reflection Within:** turns the war inward — showing how every conflict in the outer world mirrors the blindness of the mind.**Part IV – The Modern Mirror:** connects the archetype to today's politics, corporations, education, and family — showing that the ancient war still continues in boardrooms, classrooms, and hearts.

By bridging mythology with moral psychology, physics with philosophy, and spirituality with social ethics, *Dhritarashtra: The Blindness Within* transforms an ancient epic into a mirror for modern leadership and life. It reveals how awareness — not authority — is the true source of vision.

The book closes with **practical awareness tools**, including *The Awareness Audit*, *Consequence Mapping*, and *Ripple Leadership Model*, designed to help readers detect personal and organizational blindness.Through symbolic insight and scientific reasoning, it invites the reader to shift from reaction to reflection, from control to coherence, and from blindness to sight.

"Eyes may open at birth; sight opens at consequence."

In a world drowning in information but starving for wisdom, this book offers clarity.It is a reminder that awareness is not knowledge — it is the **energy that turns knowledge into vision.**

Read more at https://www.sandeepjchavan.com/.

www.ingramcontent.com/pod-product-compliance
Lightning Source LLC
LaVergne TN
LVHW090355210926
848480LV00002B/349

* 9 7 9 8 2 3 2 0 9 3 0 7 5 *